ANAEROBIC CO-DIGESTION

PRINCIPLES AND GUIDELINES TOWARD MORE EFFICIENT SYSTEMS

2025

Water Environment Federation
601 Wythe Street
Alexandria, VA 22314–1994 USA
https://www.wef.org

ISBN 978-1-57278-481-9

About WEF

The Water Environment Federation (WEF) is a not-for-profit technical and educational organization of 30,000 individual members and 75 affiliated Member Associations representing water quality professionals around the world. Since 1928, WEF and its members have protected public health and the environment. As a global water sector leader, our mission is to connect water professionals; enrich the expertise of water professionals; increase the awareness of the impact and value of water; and provide a platform for water sector innovation. To learn more, visit www.wef.org.

Prepared by the **Anaerobic Co-digestion** Task Force of the **Water Environment Federation**

Matt Higgins, *Chair*
Juliet Ohemeng-Ntiamoah, PhD, PE, *Vice-Chair*

Eider Alvarez-Puras, PE, PMP
Yamrot Amha, PhD, PE
Maxwell Armenta, PE
Vanessa Borkowski, PE
Chris Buckley
Joerg Blischke, PE
Joseph Cantwell, PE
Jayesh Carthal
Alicia Chakrabarti, PE
William Chen, PE, BCEE
Stan Chilson, PE
S. Rao Chitikela, PhD, PE, P.Eng, BCEE, F.ASCE
Tracy Chouinard, PhD, PE
Joe Cleary
Christian Davies-Venn, PhD, PE, BCEE, BC.WRE, F.ASCE, WEF Fellow
Shirin Estahbanati, PhD, PE
Richard Finger
Temesgen Fitamo, PhD, PE
Robert (Bob) Forbes, PE, MSEE
Val S. Frenkel, PhD, PE, D.WRE
Steven Greseth, MBA, PE
Felipe Gutierrez, PhD, PE
Ed Horai
Craig Just, PhD
Jonathan Koch
Erik Larson, PE
Sydney Le Cras
Xiaojin (Jim) Li
Zhongtian Li, PhD, PE
Ding-Hsiang (Shawn) Lin
Yanjin Liu
Dave Mason, WEF Fellow
Michaela McKenzie, PE
Christopher Muller, PhD, PE
Gary W. Neun
Behrang Pakzadeh, PhD, PE
Pedro Puente
Arifur Rahman, PhD, PE
William Rehkop, PE
Keval Satra, PE
Jamaie Scott, PE
Toshio Shimada, PhD, PE
Natalie Sierra, PE
Michael K. Stenstrom, PhD, PE, BCEE
Kyle Stern, PE
Bhargavi Subramanian, PhD
Rahul Subramanian, PE
Jay Surti
Jack Sutton
Christian Tasser, PE
Leila Tootchi, P.Eng
Matt Van Horne, PE, Assoc. DBIA
Caroline Wilson, PE
Randall A. Wirtz, PhD, PE
Harry Zhang, PhD, PE

Under the Direction of the **Residuals and Biosolids Subcommittee** of the **Technical Practice and Publications Advisory Panel**

Special Publications of the Water Environment Federation

The WEF Technical Practice and Publications Advisory Panel (formerly the Technical Practice Committee and the Committee on Sewage and Industrial Wastes Practice of the Federation of Sewage and Industrial Wastes Associations) was created by the Federation Board of Control on October 11, 1941. The primary function of the Committee is to originate and produce, through appropriate subcommittees, publications dealing with technical aspects of the broad interests of the Federation. These publications are intended to provide background information through a review of technical practices and detailed procedures that research and experience have shown to be functional and practical.

Contents

List of Figures

List of Tables

Preface

This technical publication aims to provide comprehensive guidelines and insights for the planning, design, operation, and maintenance of anaerobic co-digestion systems. The practice of co-digestion of wastewater residuals with high-strength organic wastes and harnessing the resulting biogas has been adopted by a growing number of water resource recovery utilities, particularly over the last decade. This technical publication intends to equip engineering and design professionals, researchers, operations and maintenance staff, educators, and students with a detailed understanding of fundamental concepts and practical applications of anaerobic co-digestion.

It should be noted that while co-digestion is practiced in sectors such as agriculture and other industrial settings, this publication is heavily focused on co-digestion at municipal water resource recovery facilities. Thus, high-strength wastes such as food waste and fats, oils, and grease (FOG) that are commonly co-digested with municipal sludge are discussed more.

The publication brings together practical experiences and research perspectives of more than 30 authors drawn from academia, research institutions, engineering consulting, municipal utilities, and technology providers. It is arranged into nine chapters covering a wide range of topics including co-digestion program development and business case evaluation, feedstocks evaluation and selection, feedstock receiving and handling, dewatering and side stream impacts, system design considerations, and biogas production and utilization. In addition, the last chapter is dedicated to case studies of full-scale co-digestion facilities in the United States where practical factors contributing to their successes and other meaningful insights are documented. Through these case studies, readers will learn the management and operational considerations needed to harness maximum benefits of co-digestion while mitigating negative impacts.

To the best of the authors' knowledge, this is the first book of its kind to provide comprehensive industry guidelines for starting and implementing co-digestion programs and will serve as an authoritative source of reference.

We express our sincere thanks to our esteemed contributors, reviewers, and WEF for making this publication possible.

Authors' and reviewers' efforts were supported by the following organizations:

Black & Veatch
Brown and Caldwell
Carollo Engineers Inc.
City of Muscatine, Iowa
Downers Grove Sanitary District, Downers Grove, IL
East Bay Municipal Utility District

First Environment
Geosyntec Consultants, Inc.
GHD, Inc.
Glenbard Wastewater Authority, Glen Ellyn, IL
Hazen and Sawyer
HR Green, Inc.
Jacobs
Johnson Controls, Inc.
Kennedy Jenks Consultants
Kiewit Corporation
Kishwaukee Water Reclamation District, Dekalb, IL
Leidos Engineering, LLC
Los Angeles County Sanitation Districts
Metro Vancouver
RC-WEE Solutions LLC
Stan Chilson LLC
Stantec Consulting Services, Inc.
Texas State University
TYLin International
University of California, Los Angeles
University of Iowa
Vaughan Company, Inc.
Veolia WTS
Water Research Foundation

Chapter 1

Introduction

Tracy Chouinard, PhD, PE, & Eider Alvarez-Puras, PE, PMP

1.0 COMMONLY USED TERMINOLOGY

This section provides some of the commonly used terminology in this publication. The most important concept to grasp before diving into all others is the differences between aerobic and anaerobic digestion. The main difference is that aerobic digestion requires oxygen, while anaerobic digestion does not use oxygen in the breakdown of organic material. Additionally, aerobic digestion is an energy-intensive process, while anaerobic digestion is less energy intensive and its byproduct, biogas, can produce energy.

Biogas: Group of gases produced by biological anaerobic degradation (decomposition in the absence of oxygen) of organic material. Methane and carbon dioxide are the primary components of biogas, with other trace gases such as hydrogen sulfide, ammonia, and hydrogen. Depending on the source of the organic material, biogas could be further categorized into digester gas, landfill gas, or manure gas, among others.

Biomethane: Upgraded form of biogas, consisting of almost 100% methane and approximately equal to natural gas in quality. Biomethane is a biogas from which carbon dioxide, hydrogen sulfide, water, and other contaminants have been removed through a purification process. Also called renewable natural gas (RNG).

Biosolids: Biosolids are produced from sewage sludge during treatment of wastewater. Biosolids are chemically, physically, and/or thermally treated sludge that are nutrient-rich solids that can be used as fertilizers or land applied according to local and federal regulations and level of treatment. Biosolids and solids are typically used interchangeably even if the biosolids are not land applied.

Blending: The action of mixing or combining two or more feedstocks together.

Co-digestion: Simultaneous anaerobic digestion of multiple organic wastes in a single digestion process unit. This process involves treating external energy-rich organic waste materials (i.e., fat, oil, and grease [FOG]; energy crops; or food waste) along with manure or wastewater sludge in digesters with excess capacity. Digestion facilities use co-digestion to increase biogas production while diverting waste from sewer collection systems or landfills.

Co-substrate: Feedstock used in co-digestion. Feedstock that is being simultaneously processed at water resource recovery facilities (WRRFs) digesters.

Digester Gas: A type of biogas specifically the product of anaerobic digestion of wastewater residuals and other feedstocks at WRRF. Also called sludge gas, sewage gas, or biogas.

Feedstocks: Any substrate, liquid or solid, that can be converted to methane by anaerobic bacteria. Feedstock can range from easily biodegradable waste to complex high-solid waste. The standard digestion feedstock at WRRF is wastewater solids. Typical co-digestion feedstocks at WRRFs are external energy-rich organic waste materials such as FOG, food and beverage processing byproducts, sugar industry slurries, food scraps, and slaughterhouse waste, among others.

FOG: Stands for fat, oil, and grease, a typical feedstock used in co-digestion programs at WRRF. These substances are byproducts of food preparation and processing, commonly originating from residential kitchens, restaurants, and industrial food processing facilities.

Sewage sludge: The solid or semi-solid product created when wastewater is treated. Sewage sludge is also known as solids or sludge if it has not been stabilized.

2.0 GENERAL OVERVIEW OF ANAEROBIC DIGESTION

Anaerobic digestion has a long history with anecdotal evidence of biogas use for heating as far back as the 10th century AD. However, the industrialization of anaerobic digestion was noted in 1859 in Bombay, India, with the first anaerobic digestion facility built at a leper colony. Later, in 1895, aerobic digestion was implemented at a wastewater treatment facility in Exeter, England, for biogas recovery to fuel streetlamps. With these first few industrialized uses of anaerobic digestion, additional research and beneficial use of the energy-rich biogas became more common especially during World War II as petroleum became scarce (National Renewable Energy Laboratory, 1998). By the 1970s, agricultural anaerobic digester technology gained momentum worldwide as oil prices continued to increase. By the latter half of the 1980s, the trend shifted from agricultural to industrial and urban waste treatment application of anaerobic digestion technologies (Bond & Templeton, 2011).

By 2020, with over 2,300 sites producing biogas in the United States, there were 520 on-farm digesters, 1,100 WRRFs with digesters, and 110 stand-alone food waste digesters, not including the landfill gas projects (American Biogas Council, 2023). As the anaerobic digestion technology advanced over the years, research and advancement into co-digestion has also progressed. These advancements included understanding the benefits of anaerobic co-digestion. Co-digestion is the simultaneous anaerobic digestion of multiple organic wastes in one digester. This chapter and subsequent chapters will further describe the synergistic benefits of co-digestion as well as the economic, operational, and maintenance benefits and challenges associated with this practice at WRRFs.

2.1 Fundamentals of Anaerobic Digestion

Anaerobic digestion is a well-established solids stabilization process. It is used extensively in WRRFs, typically on a combination of primary and secondary sludge. It may also be used in larger facilities without primary clarification. Smaller facilities without primary clarification tend to use aerobic digestion, due to its lower capital cost and lower operational complexity. During anaerobic digestion, the biodegradable organic matter in the sewage sludge is converted to biogas and residual matter, thus reducing the mass and volume of wastewater solids, stabilizing the solids by reducing pathogens, and producing biogas that can be used to help offset energy demands. When evaluating anaerobic digestion as the biosolids stabilization process at a WRRF, special consideration must be given to the life-cycle costs (economic and operational considerations), the biosolids volume and desired use (environmental considerations), the ability to sustainably reuse the biosolids and biogas products (sustainability considerations), as well as the socio-political impact in the community.

2.1.1 Biochemical Reactions

Anaerobic digestion is the biological process that degrades organic matter in the absence of oxygen in a series of microbiological pathways. These series of reactions are commonly considered a four-stage process as shown in Figure 1.1:

1. Hydrolysis—Solubilization of particulate organic compounds (such as carbohydrates, fats, and proteins) into simple soluble compounds.

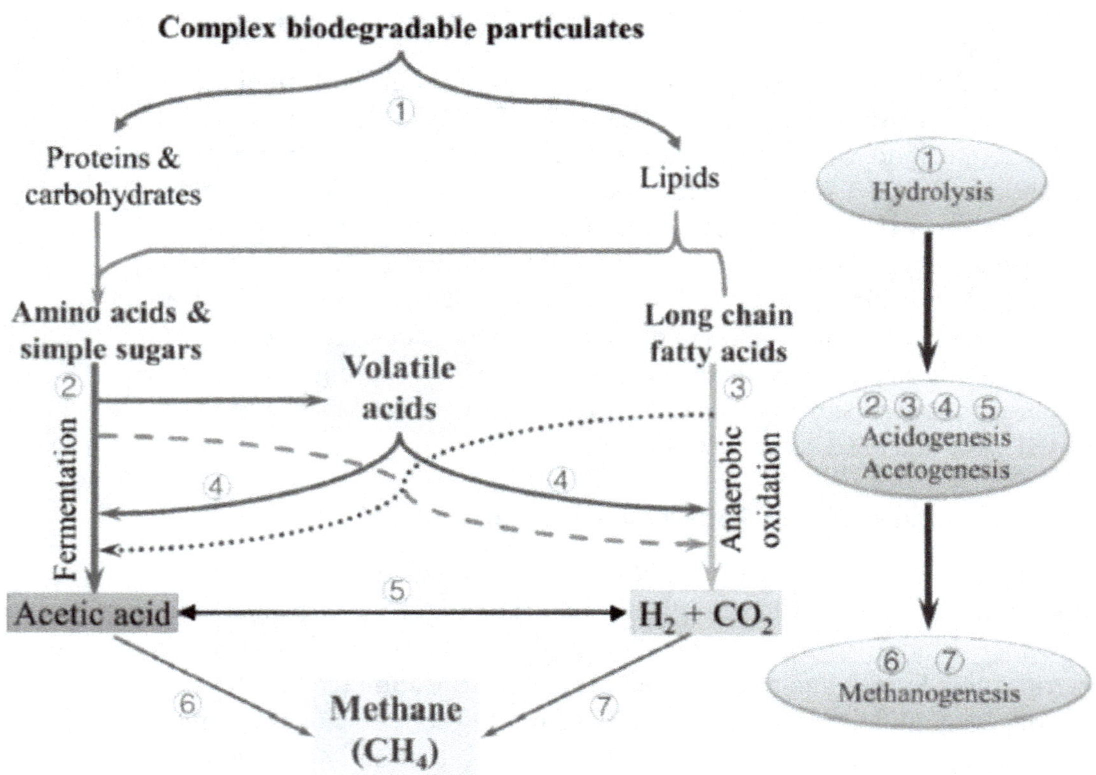

FIGURE 1.1 Anaerobic Digestion Biochemical Pathways (Adapted from WEF, 2021)

2. Fermentation (or Acidogenesis)—Simple soluble compounds (such as sugar, amino acids, fatty acids) are converted to volatile acids and alcohols.
3. Acetogenesis—Conversion of volatile acids and alcohols to simpler substrates such as acetic acid or acetate and hydrogen gas.
4. Methanogenesis—Production of methane, as shown in Figure 1.1, occurs through the degradation of acetate into methane and carbon dioxide (6) and the reduction of carbon dioxide utilizing hydrogen (7).

2.1.2 Microbial Processes

The anaerobic digestion process provides a balanced ecological environment for microorganisms to work interactively and interdependently to successively break down complex biodegradable particulates into biogas and residual organic matter (also called digestate). The microbial diversity in the anaerobic digesters depends on various factors such as feedstock type, seed inoculum, temperature, granulation, mixing intensity, pretreatment type, digester design, organic loading rate, solids retention time, and hydraulic retention time. Archaea, the major microorganisms involved in methanogenesis, are sensitive to the various environmental parameters (Harirchi et al., 2022), such as pH, ammonium concentration, and volatile acids concentration, with acetoclastic methanogens being more sensitive than hydrogenotrophic methanogens. Implementing a co-digestion program at an existing anaerobic digestion facility by feeding external waste into the digesters will modify the digester's operating parameters, and consequently, its microbial diversity and process performance. Characterization of supplemental feedstocks, adjustment of digester operating parameters, and digestion performance monitoring are critical for a successful implementation. Chapter 3 provides more details on this topic.

2.1.3 Different Types of Anaerobic Digestion

Digesters can be classified as wet and dry, depending on the moisture content of the feedstock. Wet digesters, also called low-solids anaerobic digesters, are the most common at WRRFs. Wet digesters generally process feedstock with less than 15% solids content in a slurry form that can be more easily mixed and pumped. The typical slurry is less than 5% solids, as mixing and pumping sludge above 8% solids becomes challenging. Wet digesters are typically complete-mix reactors that are continuously fed but can also be fed in batch mode. Dry digesters, also called high solids anaerobic digesters, typically process feedstock with greater than 15% solids content in plug-flow-type vessels. These reactors are typically used in the solid waste industry and are not common at WRRFs. Also, these technologies are primarily installed in Europe and have limited mixing options (Water Environment Federation [WEF], 2023).

This publication is focused on wet digestion at WRRFs unless specifically mentioned otherwise.

2.2 Typical Types of Substrates for Co-digestion

Due to digestion systems typically operating below rated capacity, it is common for WRRFs to have extra capacity in their anaerobic digestion systems. An increasing number of WRRFs are looking to utilize the spare digestion capacity to treat biodegradable external organic waste, by implementing a co-digestion process. Co-digestion consists of simultaneously processing WRRF solids with external hauled-in organic waste in anaerobic digesters. Many European and North American WRRFs have implemented co-digestion programs where they digest wastewater sludge with other organic waste, such as FOG, food scraps, sugar processing waste, or slaughterhouse slurries.

Feedstocks utilized in co-digestion, also called co-substrates, vary considerably in composition, homogeneity, fluid dynamics, biodegradability, and methane potential. Chapter 3 will provide detailed information about the typical types of substrates used for co-digestion. These substrates can have different effects on biogas production and composition, byproducts, ancillary processes, and side stream impact.

3.0 TYPICAL DRIVERS ASSOCIATED WITH CO-DIGESTION

Implementing a co-digestion program at a WRRF can depend on several drivers such as regulatory changes, utilizing existing systems, improved economics, and environmental and social governance goals. The following are examples of co-digestion programs implemented at WRRFs and the drivers:

- Maximizing underutilized WRRF assets
- Increasing biogas production to optimize the economics of biogas utilization projects
- Diverting challenging waste (i.e., FOG) from collection systems to reduce maintenance costs
- Diverting waste from landfills to extend their life expectancy
- Creating a revenue stream from the received organic waste hauled-in tipping fees

- Creating a revenue stream from the sale of biogas utilization products (i.e., in-house energy needs offset, exported electricity, sale of biogas rights to third parties, renewable natural gas sale)
- Improving the local sociopolitical environment by addressing waste management strategies, renewable energy forms, greenhouse gas reduction, and circular economy.

This section provides a high-level introduction to these drivers. These will be outlined in further detail in subsequent chapters.

3.1 Regulatory Changes

One of the key regulatory changes that can drive a WRRF to start a co-digestion program is restrictions or ban of organic waste in landfills. These bans divert organic waste, typically food waste, from the landfill, requiring commercial producers to seek new outlets for their wastes such as anaerobic digesters or composters. States such as California, New Jersey, Washington, Massachusetts, and Minnesota, to name a few, have bans or reduction laws that drive co-digestion. In addition to the benefits that WRRFs may achieve with these co-digestion programs, diversion of organic material from the landfill to an anaerobic digester reduces the amount of methane that is released from a landfill, as these organic materials break down. In a digester, the breakdown of organic material to methane (and biogas) is more easily captured and utilized than in a landfill.

3.2 Maximize Existing Assets

When evaluating a co-digestion program, WRRFs with underutilized systems may benefit more from a co-digestion program as minimal new infrastructure needs to be constructed. For instance, if a WRRF has additional existing capacity in digesters, this capacity can be used by a hauled-in organic waste. The additional capital expenditure to implement an organics program would be less than if the facility had to construct new anaerobic digesters.

3.3 Economic Impacts

Co-digestion programs are driven by the WRRF's desire to fully utilize its existing assets and maximize its monetization. The monetization of anaerobic digestion byproducts results from two revenue streams: tipping fees from the received feedstocks, and energy savings or revenue generation from biogas utilization projects. Co-digestion programs enhance biogas production while receiving tipping fees, improving the feasibility of biogas-to-energy projects due to the economies of scale. With energy representing one of the largest WRRF operational costs and biogas and biomethane being the only green alternatives to natural gas, co-digestion presents a pivotal opportunity to maximize biogas monetization.

3.3.1 Revenue Potential

To perform a comprehensive economic evaluation of a co-digestion program, the volatility in the monetization of the biogas-end-product, the co-substrate tipping fees, and biosolids management fees must all be considered. There are multiple types of biogas utilization projects. The majority of the biogas use applications involve generating electricity, heat, or both. Due to recent global efforts to decarbonize the natural gas supply, converting digester gas into renewable natural gas, for injection into vehicles or a pipeline, has been gaining increased traction.

Historically, WRRFs have utilized biogas as digester's hot water boiler fuel to offset in-house thermal energy requirements. This is the simplest use because the biogas does not typically have to be conditioned to remove contaminants, as boilers have a relatively high tolerance compared to other biogas combusting equipment. Due to the seasonal variability of thermal demands, it is typical for WRRFs with biogas-to-boiler utilization to flare excess biogas.

The second most common use of biogas is as fuel to produce electrical power (biogas-to-electricity) or combined heat and power (biogas-to-CHP), also known as cogeneration. This biogas application includes technologies such as internal combustion engines, combustion gas turbines, microturbines, steam boilers with steam turbine generators, fuel cells, and, more recently, linear motion generators. These applications typically require some level of biogas conditioning to remove gas contaminants (i.e., hydrogen sulfide and silica compounds) and protect the integrity of the equipment.

In the cases of biogas-to-thermal and to-electricity, the savings or revenue generation is tied to the utility rate, for energy purchase or electricity export, set by the local utility for that particular WRRF. Electrical rates have historically been steadily increasing, with natural gas rates being more volatile than electricity due to natural gas prices being susceptible to global market trends. Due to energy rates and policies differing substantially per state, local market, and WRRF user, it is preferable to perform a site-specific economic feasibility evaluation.

A biogas utilization application recently gaining traction is biogas-to-renewable natural gas (biogas-to-RNG). Recent high market prices for commercial natural gas and increased decarbonization commitments from energy and private companies have made exporting green natural gas an economically feasible technology for certain WRRFs. Converting biogas to renewable natural gas for injection into vehicles or a pipeline requires a higher level of biogas conditioning than for previously mentioned applications. The renewable natural gas product, virtually indistinguishable from pipeline natural gas, can be exported to third parties interested in purchasing this commodity or used in-house by WRRF with high vehicle fuel costs.

While the sale of the natural gas commodity is an important economic factor, optimizing the monetization of the environmental attributes associated with the renewable aspect of RNG is critical. The environmental attributes can be sold by WRRFs as renewable identification numbers (RIN) credits through the Renewable Fuel Standard federal program to obligated parties in short-term sale agreements or sold to non-regulated parties interested in decarbonizing their operations in long-term sale agreements. Co-digestion complicates RIN classification because it involves multiple feedstock that may fall under different RIN categories, requiring detailed tracking and compliance with complex regulatory requirements. While the RIN market will continue to be volatile, the recent demand increase for long-term agreements from non-obligated parties is providing a backstop on environmental attributes RIN prices, de-risking the monetization for this particular biogas application.

3.3.2 Management Costs

Other costs that can drive co-digestion programs are biosolids management costs. These are the hauling and tipping fees the WRRFs will pay for biosolids management. Depending on how the biosolids are managed and the location of the WRRFs, hauling and tipping fees can range from $30 to $200 per wet ton. Higher management fees will drive WRRFs to seek organic feedstocks that have lower solids contents, such as FOG, spoiled sugar water, or whey, while reducing the desire for feedstocks such as food waste, which tends to be higher in solids content.

3.4 Sustainability Goals

Many municipalities, utilities, and WRRFs have sustainability goals that they would like to achieve. These can be goals such as a reduction in greenhouse gas emissions, becoming net energy neutral, or even producing a surplus of clean energy. Measurement and evaluation of anaerobic digestion for these goals can be completed using the United Nations Sustainability Development Goals (SDG) framework, which is an internationally accepted framework with defined goals until 2030. Under SDG, there are 17 clearly defined goals with 169 targets and 231 detailed indicators (Piadeh et al., 2024). The 17 goals are as follows: no poverty; zero hunger;

good health and well-being; quality education; clean water and sanitation; affordable and clean energy; decent work and economic growth; industry, innovation, and infrastructure; reduced inequalities; sustainable cities and communities; responsible consumption and production; climate action; life below water; life on land; peace, justice, and strong institution; and partnerships for the goals (Obaideen et al., 2022; Piadeh et al., 2024). Based on these goals and the attributes of biogas, Obaideen et al. (2022) noted that anaerobic digestion directly impacts 12 of the 17 goals, while Piadeh et al. (2024) indicated that anaerobic digestion can contribute to all 17 goals, but the authors recognize that the contributions to the social goals are only indirect. As noted above, a few of the goals can be directly contributed to co-digestion such as production of clean energy, economic growth, clean water and sanitation, and responsible consumption and production, to name a few. As WRRFs evaluate co-digestion programs, they can use the SDG framework to aid in measuring and developing achievable and attainable goals. This framework can also show and support how co-digestion supports a circular economy with the organic material being converted to clean energy and nutrient soil amendments, to name a few aspects.

4.0 OVERVIEW OF CO-DIGESTION PROGRAMS AT WATER RESOURCE RECOVERY FACILITIES

4.1 Feedstock Procurement Models

Co-digestion programs can be structured in several different ways. The case studies will discuss how those individual facilities structure their programs. For procurement of feedstocks, a WRRF can work with a third-party vendor to process the feedstocks. This is typically done with food waste. For instance, food waste would be collected from commercial facilities or source-separated from the municipal solids waste. Then, it would be processed by removing any contaminants, and then mashed and pulped into a pumpable slurry. This slurry can range from 14%–20% total solids. The vendor would then haul this slurry to the WRRF. Central Marin Sanitation District (California) has a program similar to this. Additionally, the WRRF can manage the program themselves with published tipping fees such as East Bay Municipal Utilities District in California or Greater Lawrence Sanitation District in Massachusetts. Finally, another feedstock procurement model is when the WRRF works with an energy services company. This type of company will do the market assessment to determine the types of feedstocks available as well as the tipping fee potential. They will typically guarantee the volume and tipping fee of feedstock with some kind of revenue share agreement. This company would be fully responsible to secure feedstocks coming to the facility.

4.2 Biogas Production Utilization

A key driver for implementing a co-digestion program is the increase in biogas production and associated increase in the savings or revenue from the biogas utilization. WRRFs practicing co-digestion have reported increased digester gas production ranging from 10% to more than 75% after adding offsite feedstocks (WEF et al., 2023). The biogas production increase will vary depending on the type of feedstock, the particular sludge at the WRRF, as well as the site-specific digester operations. As mentioned previously, utilization of biogas can impact revenue potential significantly at a WRRF. With a co-digestion program increasing biogas production, these revenues can make the programs more attractive as they essentially can pay for themselves.

4.3 Facility Process Potential Impacts

Implementing co-digestion will have impacts to the facility. Those impacts will vary depending on the type of feedstock(s) co-digested and the process being evaluated. In general, the addition of food waste has shown an increase in cake solids after dewatering (Higgins & Rajagopalan, 2017; Higgins et al., 2017); however, as food waste can be protein-rich, it can increase the ammonia within the system as it breaks down (Higgins & Rajagopalan, 2017; Mehariya et al., 2018). This can cause an increase in ammonia within the digester as well as the dewatered filtrate. The additional nutrient load would need to be recovered via side stream treatment or treated within the WRRFs process. Additionally, food waste tends to be high in organic content and has a high percentage of volatile solids reduction, which results in an insignificant increase in the amount of biosolids that needs to be managed (Mehariya et al., 2018).

Alternatively, FOG has a higher lipid content and lower protein content than food waste. Higgins and Rajagopalan (2017) observed at higher organic loading of FOG there was a decrease in ammonia. Silvestre et al. (2014) indicated that the higher levels of long-chain fatty acids also caused decreased dewaterability and instability in thermophilic digested sludge. As Higgins and Rajagopalan (2017) indicated, the dewaterability performance of co-digested sludge correlated to the monovalent/divalent ratio. As this ratio increased (more sludge with monovalent bonds such as sodium [Na^+] and potassium [K^+] relative to divalent bonds such as calcium [Ca^{2+}] and magnesium [Mg^{2+}]), the cake solids decreased.

Further details on impacts of co-digestion in a facility and its processes will be provided in subsequent chapters.

5.0 REFERENCES

American Biogas Council. (2023). *Biogas market snapshot.* https://americanbiogascouncil.org/biogas-market-snapshot/

Bond, T., & Templeton, M. R. (2011). History and future of domestic biogas plants in the developing world. *Energy for Sustainable Development, 15*(4), 347–354.

Harirchi, S., Wainaina, S., Sar, T., Nojoumi, S. A., Parchami, M., Parchami, M., Varjani, S., Khanal, S. K., Wong, J., Awasthi, M. K., & Taherzadeh, M. J. (2022). Microbiological insights into anaerobic digestion for biogas, hydrogen or volatile fatty acids (VFAs): A review. *Bioengineered, 13*(3), 6521–6557.

Higgins, M., & Rajagopalan, G. (2017). Understanding impacts of co-digestion: Digester chemistry, gas production, dewaterability, solids production, cake quality, and economics. Water Environment Research Foundation. Contract No. ENER12R13.

Higgins, M., Rajagopalan, G., Miller, A., Brown, J., & Beightol, S. (2017). Effect of food waste co-digestion on digestion, dewatering, and cake quality. *Water Environment Research, 89*(1), 32–42.

Karki, R., Chuenchart, W., Surendra, K. C., Shrestha, S., Raskin, L., Sung, S., Hashimoto, A., & Khanal, S. K. (2021). Anaerobic co-digestion: Current status and perspectives. *Bioresource Technology, 330*, 125001.

Mehariya, S., Patel, A. K., Obulisamy, P. K., Punniyakotti, E., & Wong, J. W. (2018). Co-digestion of food waste and sewage sludge for methane production: Current status and perspective. *Bioresource Technology, 265*, 519–531.

National Renewable Energy Laboratory. (1998). *Methane recovery from animal manures: The current opportunities casebook* (pp. 17–20). https://www.nrel.gov/docs/fy99osti/25145.pdf

Obaideen, K., Abdelkareem, M. A., Wilberforce, T., Elsaid, K., Sayed, E. T., Maghrabie, H. M., & Olabi, A. G. (2022). Biogas role in achievement of the sustainable development goals: Evaluation, challenges, and guidelines. *Journal of the Taiwan Institute of Chemical Engineers, 131*, 104207.

Piadeh, F., Offie, I., Behzadian, K., Rizzuto, J. P., Bywater, A., Córdoba-Pachón, J. R., & Walker, M. (2024). A critical review for the impact of anaerobic digestion on the sustainable development goals. *Journal of Environmental Management, 349*, 119458.

Silvestre, G., Illa, J., Fernández, B., & Bonmatí, A. (2014). Thermophilic anaerobic co-digestion of sewage sludge with grease waste: Effect of long chain fatty acids in the methane yield and its dewatering properties. *Applied Energy*, *117*, 87–94.

Water Environment Federation. (2021). *Energy in water resource recovery facilities* (2nd ed.; Manual of Practice No. 32).

Water Environment Federation. (2023). *Standards for design of water resource recovery facilities* (WEF 8).

Chapter 2

Co-digestion Program Development and Business Case Evaluation

Maxwell Armenta, PE; Alicia Chakrabarti, PE; Christopher Muller, PhD, PE; William Rehkop, PE; & Natalie Sierra, PE

1.0 PROGRAM PLANNING AND PERMITTING

There are a number of initial questions to consider in program planning. The first is whether there is excess capacity, or if additional capacity can be created through operational changes. A co-digestion program will rarely be economically feasible in the absence of existing excess capacity. Other basic considerations include land to route truck traffic and install a waste receiving station; availability of co-digestion wastes; impacts to existing permits; and new permitting requirements.

Once an agency has determined that co-digestion is viable, program planning can begin. This includes assessing regulatory requirements and setting up the administrative elements to permit, track, monitor, bill, and receive wastes from the waste generator and/or the hauler. Figure 2.1 provides an overview of potential impacts from a co-digestion program on the water resource recovery facility (WRRF) and the larger community.

1.1 Co-digestion Feasibility

Figure 2.2 shows a process flow diagram incorporating questions that should be asked and answered to determine whether a co-digestion program is feasible. Key considerations for co-digestion feasibility include the following elements:

- **Feed stock availability**—evaluate whether wastes are available for co-digestion at the WRRF. This can include a market assessment to interview grease haulers, solids waste companies, local municipalities, or industries.
- **Capacity**—evaluate the capacity of anaerobic digestion, dewatering, biosolids management, and biogas utilization. If significant quantities of co-digestion waste are received, then evaluating additional nutrient loading may be considered. Capacity evaluations, such as digester residence time and volatile organic loading, are discussed in subsequent chapters.

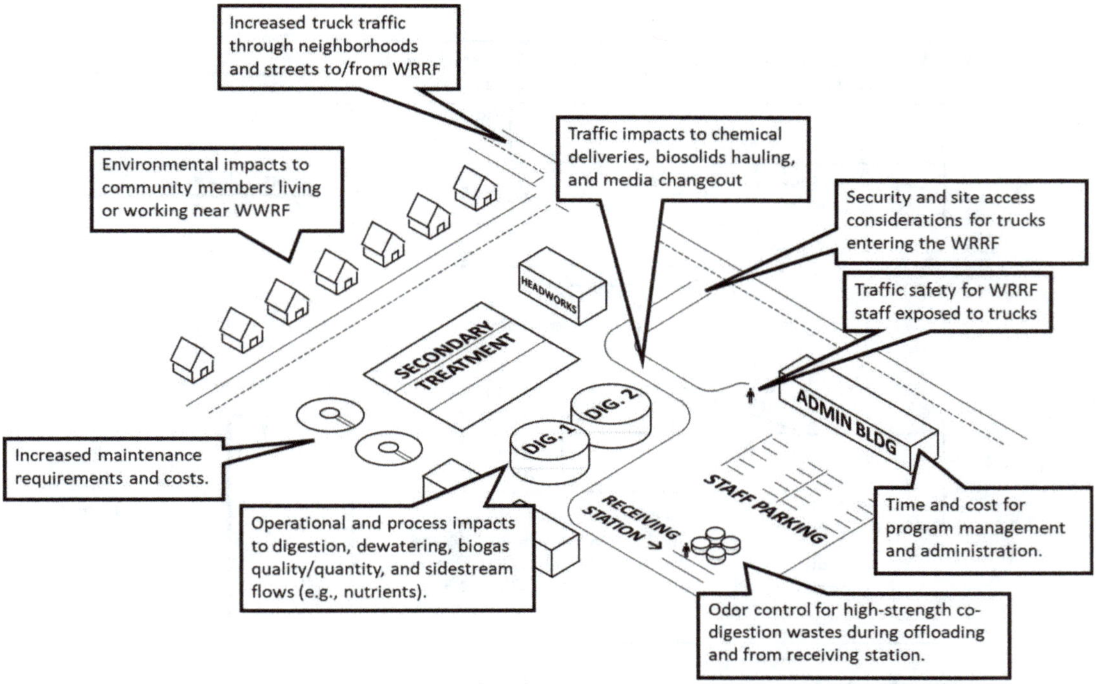

FIGURE 2.1 Co-digestion Impacts on the Water Resource Recovery Facility, Operations, and Larger Community

- **Regulations**—evaluate whether there are significant regulatory hurdles. Waste can impact existing regulatory requirements from the WRRF due to changes in biogas, biosolids, or final effluent quality. The changes may impact air quality, solid waste, or national pollutant discharge elimination system (NPDES) permits, respectively. Implementing a co-digestion program may require development of program administration documentation or revision of the facility description in the NPDES permit. Additional compliance may be required for certain waste types.
- **Environmental justice**—evaluate the impact of increased truck traffic (e.g., fuel emissions, noise pollution, increased roadway wear) and offloading (e.g., odors, spillage) to the community near the WRRF. A co-digestion program could negatively impact neighbors and staff without proper program development and facility design.
- **Business case evaluation**—evaluate the economic impact of co-digestion to demonstrate a return on investment (ROI) such that rate payers are not subsidizing a co-digestion program. In many cases, a business case evaluation may demonstrate co-digestion as a means to

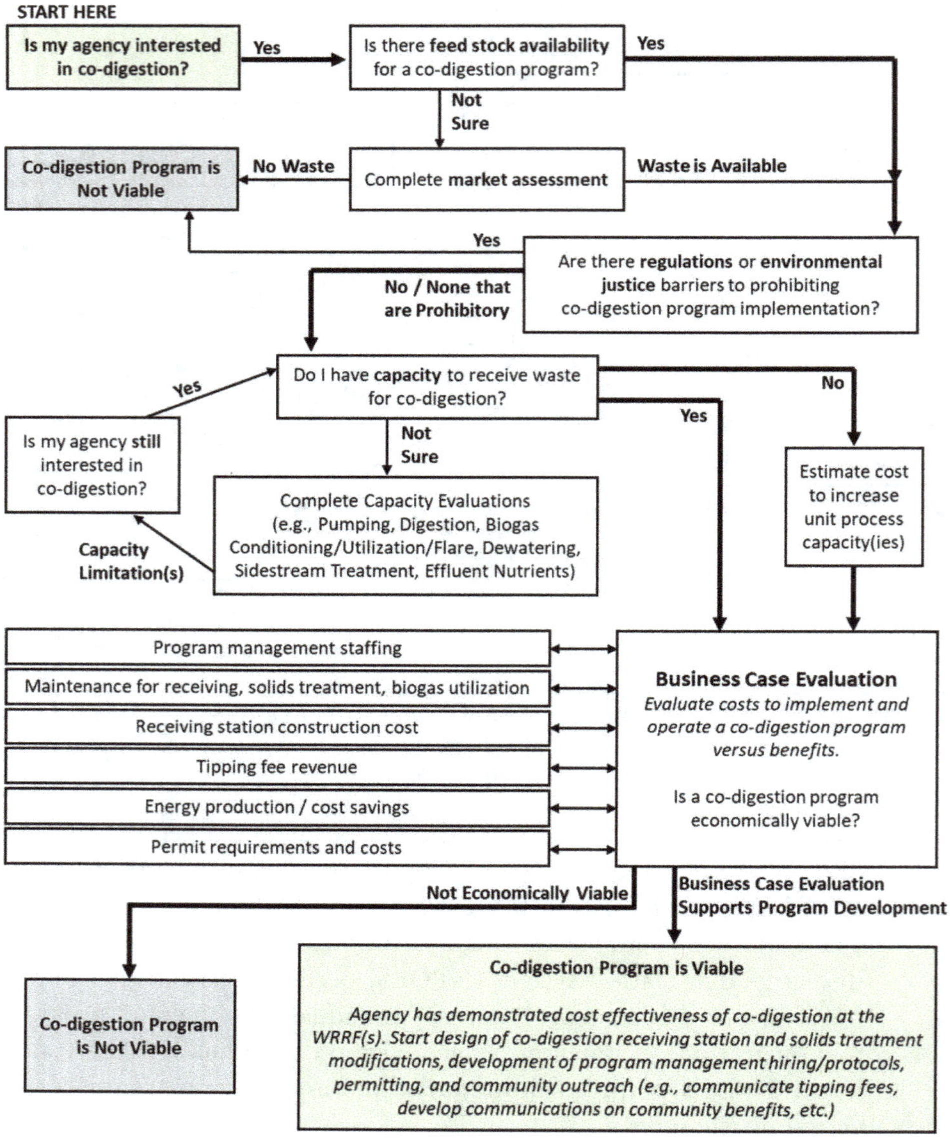

FIGURE 2.2 Program Planning Decision Making Process Flow Diagram

maintain lower rates. A triple bottom line (TBL) evaluation can be incorporated to evaluate community benefits from co-digestion.

1.1.1 Alignment or Disconnect Between Co-digestion and Water Resource Recovery Facilities Goals

Before implementing a co-digestion program, the WRRF needs to evaluate the benefits and challenges of co-digestion. Alignment between co-digestion impacts and the WRRF goals is necessary for program acceptance and long-term success. This includes buy-in from public officials, board members, management, engineering, operations, maintenance, and laboratory staff. Co-digestion provides many benefits to a WRRF, the surrounding community, and the world. Tipping fee revenues and increased biogas can provide energy resilience and supplement rates to address capital improvement program (CIP) costs. A co-digestion waste receiving station can provide a benefit to local industries by providing a local, cost-competitive source for their waste. Waste streams used for co-digestion and energy production decrease fossil fuel use at the WRRF and help with climate action plan goals.

Co-digestion also presents challenges and drawbacks. Increased truck traffic may be a safety concern or annoyance to the WRRF staff and surrounding communities. Co-digestion odors can be an issue for sensitive receptors if facilities are not properly designed or if offloading is not properly managed. Program management and monitoring require additional systems to track waste haulers and may require increased staffing. Increased maintenance and operating requirements can be viewed as an unnecessary headache for WRRF staff, or may cause significant operational issues such as digester upsets (discussed in subsequent chapters). The additional staffing and maintenance requirements add to annual operating costs.

Decision makers should document primary reasons for pursuing a co-digestion program. If there is a disconnect between co-digestion benefits and the WRRF's goals, then challenges associated with co-digestion may not be worth the investment. Priorities for the WRRF can be reconsidered with time. A co-digestion program may not currently make sense for the WRRF, but changes to economic factors or organization priorities may make a program viable in the future.

1.1.2 Market Conditions: Waste Availability, Quality, and Competition

The decision to implement a co-digestion program should be backed by a business case evaluation or TBL evaluation, both of which will be based on assumed co-digestion waste receiving and WRRF performance. A fundamental understanding of local market conditions is needed at the inception and continued execution of a co-digestion program. Understanding market conditions improves assumptions on waste availability and economic risk. The following factors should be included in a market assessment:

- **Current management practices of potential co-digestion waste sources**—most organics are managed through the private sector with existing waste management strategies/infrastructure. A new waste management option offered through a co-digestion program will be considered by businesses based on location, access, and cost (i.e., does co-digestion improve their operating margins). A survey of local industries (e.g., grease trap haulers, solids waste management companies) should be conducted to evaluate their current management practices and operational costs.

- **Quality of potential co-digestion waste sources**—common co-digestion wastes such as fats, oils, and grease (FOG) and food waste have characteristics that are well documented, but surveying/sampling local industries will provide a better estimate of WRRF impacts and co-digestion performance. Better predictions can result in a more accurate economic evaluation and inform receiving station design.
- **Competition**—competition for preferred waste types (easy to accept and pump, low WRRF process impacts) from the private sector and/or other municipalities can impact the quantity of waste received and influent tipping fee decisions. Understanding the proximity, capacity, and tipping fees of competing facilities will provide a more accurate understanding of co-digestion program payback and the relative financial risk associated with a potential reduction in anticipated revenues.
- **Business practices and goals**—the emergence of zero-waste sustainability goals for different business sectors can present opportunities for co-digestion. Outreach to local businesses can identify synergies that anchor a new program with guaranteed quantities of feedstocks from specific industry partner(s).

There are several other considerations surrounding the development of a co-digestion program that may impact its long-term viability or change the outcome of a co-digestion feasibility study. Co-digestion program plans should account for uncertainty. For example, some WRRFs have reported greater quantities of co-digestion waste than anticipated, whereas other WRRFs reported initial success but competition from other co-digestion facilities that required additional effort to keep attracting waste haulers. Ultimately, a greater understanding of current and ongoing market conditions provides a greater probability of co-digestion program success.

1.2 Regulatory Requirements and Drivers

The regulatory framework around co-digestion relates to two major aspects: (1) the receipt of co-digestion wastes, which will typically fall under a state's solid waste regulations (in the United States), and (2) the management of the resulting anaerobic digestion products (i.e., biogas, biosolids, and dewatering return). An increase in biogas generation may impact air permit compliance. Changes in biosolids quality and/or quantity can impact biosolids management, and dewatering return streams can impact effluent nutrient concentrations, which may impact compliance with an NPDES permit. It is also important to note that in instances where food waste only digestion is occurring, different state regulations and permits may be triggered than those discussed below for co-digestion. The WRRF's own charters and rules need to also be considered.

In addition to regulatory requirements that may challenge co-digestion program implementation, regional regulations around solid waste can incentivize co-digestion. Governments with regulations requiring organic waste diversion from landfills have resulted in a relatively high number of co-digestion facilities, such as California, Maine, and New York in the United States and in Germany (Rauch-Williams & Schaum, 2018). Renewable energy targets and organic waste diversion have also incentivized co-digestion globally (Global Methane Initiative, 2014).

1.2.1 National Pollutant Discharge Elimination System

Co-digestion can increase nutrients in final effluent due to dissolved nitrogen and phosphorus released in digestion that are recycled to liquid treatment from the dewatering process. Depending on the quantity and quality of waste received, this can impact NPDES permit compliance. Individual WRRFs should also coordinate with their designated NPDES permitting authority to determine whether any specific changes to their permit are necessary. The permitting authority likely needs to verify that co-digestion waste will not require monitoring of new parameters outside of existing permit requirements. Given that co-digestion is a routinely accepted practice, no modifications would likely be necessary for typical waste streams.

Federal pretreatment regulations include a specific prohibition on wastewater with pH lower than 5.0, unless specifically designed to accommodate such discharges (Chapter 40 Part 503 of the *Code of Federal Regulations* [40 CFR 503] 5.b.2). Grease interceptor wastes (i.e., FOG) contain degraded oil with free fatty acids, which reduce the pH. To comply with federal pretreatment regulations, receiving stations should be designed to accept these wastes.

1.2.2 Biosolids End Use Requirements

The federal regulatory framework for biosolids regulations, 40 CFR 503, provides standards for biosolids management, maximum pollutant concentrations, requirements for vector attraction reduction, and requirements associated with pathogen reduction. Individual states and counties may have additional biosolids-related regulations, particularly associated with land application, which must also be considered. Potential changes in federal and local biosolids regulations should be considered if co-digestion wastes contain contaminants such as per- and polyfluoroalkl substances (PFAS) or other industrial byproducts.

1.2.3 Waste Receiving Requirements

Certain types of wastes, such as solid food waste, may have their own specific regulations regarding receipt and final disposition. For example, they may be regulated under state solid waste law with specific requirements for both the hauler and receiver (in this case, the WRRF). See Section 1.2.5 for an expanded discussion.

1.2.4 Air Quality Requirements

Foul air from co-digestion waste receiving stations may be regulated by local entities to limit constituents. This may require odor control equipment with routine sampling and reporting requirements for constituents such as hydrogen sulfide or volatile organic compounds. Certain wastes contain sulfur compounds that are converted to hydrogen sulfide in the digesters, which may require additional treatment or control. Co-digestion programs are typically developed to beneficially use biogas. The additional biogas that is produced from co-digestion will increase constituent loading at flares or exhausts. Air permit modifications to account for the additional biogas production and/or additional combustion capacity may be required. This could limit the amount of biogas that can be produced and/or utilized depending on the permitting agency.

1.2.5 Permitting Requirement Examples

Permit requirements are regulated at both the national and regional levels (e.g., federal and state levels for the United States). Examples from individual states within the United States are included below to demonstrate differences in permitting requirements for facilities implementing co-digestion:

- In 2015, CalRecycle, the governing body for solid waste regulations in California, promulgated new compost regulations dictating general standards and permitting requirements for new and existing composting facilities. The applicability of these regulations has set out specific definitions for composting facilities to include "facilities that receive, store, handle, recover, transfer, or process solid waste." In crafting regulations, CalRecycle created an exemption from compost regulations specifically for WRRFs incorporating a co-digestion program (but notably, not food waste; only digestion).

- Inedible Kitchen Grease (IKG) collection, transport, and rending is also regulated in California. IKG haulers are required to display a decal to demonstrate they are registered and are using proper techniques during collection and transport. WRRFs in California must also be licensed under IKG and maintain records of all grease deliveries.
- New regulations are in development that will impact codigestion at WRRFs. For example, at the time of publication, Pennsylvania's Department of Environmental Protection proposed draft changes to the general permit programs associated with biosolids management, with the potential to permit those utilities receiving high-strength wastes for co-digestion as solid waste receiving facilities.

1.3 Waste Generator and Hauler Permitting

The WRRF should consider a permit application form for waste haulers before receiving trucked waste (U.S. Environmental Protection Agency, 1999). This process allows for the WRRF to collect information on the waste generator and hauler, including characterization of waste and estimation of waste deliveries. Basic information such as the name of the business(es), contact person, phone number, email, and address should be collected for billing purposes. A permit application fee should be assigned to cover the WRRFs costs to review and administer the permit. The following elements may also be considered in the permit application process.

1.3.1 Insurance Requirements

A waste hauler permit provides the opportunity to require insurance coverage by the hauler for general, auto, and/or pollution liability. The latter provides coverage in case the hauler is responsible for environmental pollution directly from transport of waste material or indirectly from a process upset to the WRRF.

1.3.2 Vehicle Information and Identification

A waste hauler permit may include information on the waste hauler's expected and authorized vehicles that will be used to transport and discharge waste. Information may include the tank volume, license or registration information, and a vehicle identification number. Tank volume may be used for billing purposes if discharged waste flows are not directly measured. The WRRF may elect to provide adhesive identification numbers specific to the co-digestion program depending on the program management approach, or government-provided vehicle identification may be used (e.g., license plates).

1.3.3 Waste Characterization and Compliance

Before accepting co-digestion waste, characterization should be conducted to evaluate whether the waste will be compliant with regulatory requirements and not cause interference or pass through at the WRRF. The waste hauler permit may be used to specify additional waste characteristic requirements for co-digestion waste, such as a range of acceptable total and volatile solids (minimum or maximum), maximum contamination levels, or a range of acceptable pH values.

1.4 Waste Tracking and Monitoring

Either automated or manual methods may be implemented to track waste receiving. Automated check-in systems can be installed to log driver and hauler information, or staff can be located near a receiving station to document when waste haulers arrive using pen and paper logs or computers. Information such as waste hauler arrival, discharge duration, discharge volume (either via flow measurement, level measurements, or approximation based on tank size), truck weight (before and after discharge), and waste type can be tracked. This information can be used for upset troubleshooting, monitoring and compliance, billing, and process modeling for planning/optimization.

Monitoring waste hauler activity conveys to haulers that the WRRF is closely managing the program. This may disincentivize bad actors and can help minimize non-compliant waste types that may result in process upsets. Oversight by staff members at receiving stations and analysis of grab samples satisfies monitoring requirements.

1.5 Billing System

A successful co-digestion program typically requires collection of tipping fees for ROI. This requires the WRRF to use information collected on the quantity and type (and quality, if applicable) to calculate fees. Billing requires time from existing or new staff, and may require follow-up communications to receive overdue payments. The WRRF will need to determine whether new hardware or software is required for billing. Waste hauler invoicing may be integrated into the WRRF's existing accounting software. Invoices can be generated by manual entry or tied into an automated tracking system, if installed.

1.6 Receiving Station: Inclusion, Siting, and Land Availability

Receiving stations are a best practice for co-digestion programs, as discussed in Chapter 6, due to significant process benefits provided by flow equalization, blending, and steady digester feeding. A receiving station should be located close to digesters or blending tanks to minimize piping. The receiving station needs to be accessible to trucks with sufficient space for offloading. If solid wastes require treatment, source separation, or processing before feeding to digestion (e.g., food waste slurry), the WRRF should evaluate which unit processes, if any, will be located on-site and operated by WRRF staff. Subsequent chapters discuss pretreatment in greater detail.

1.7 Truck Routing and Site Access

Waste haulers will increase traffic and may impact the useful life of roadways within and outside the WRRF. Truck weight limitations on roadways in the neighboring communities need to be considered so trucks can access the WRRF. Traffic within the WRRF can interfere with regular operations, including chemical deliveries and biosolids hauling. Traffic can also interfere with construction activities and facility maintenance. Signs may be installed to direct traffic within the WRRF and to limit speed. Roadway improvements or new construction can facilitate improved traffic flow. The WRRF should also consider whether the connection point for hauler discharge will be located within or outside of the WRRF fence line. This impacts site access and security.

2.0 PROGRAM MANAGEMENT

Internal and external coordination and management are key elements of a co-digestion program. Depending on the size of the agency and/or program, additional staff may be required to administer and enforce permits, conduct billing, interact with the truck drivers, and operate/maintain receiving stations. Careful evaluation of proposed waste streams and ongoing sampling and analysis is necessary to protect WRRF operations, ensure compliance with permits, and maintain high-quality biosolids and effluent. Table 2.1 presents elements of program management and summarizes low and high effort approaches to addressing these elements.

2.1 Co-digestion Program Staffing

Staff are required to manage and administer the co-digestion program. The number of staff required depends on the quantity of waste received and the size of the WRRF. Small- and medium-sized WRRFs with modest addition of co-digestion waste have been able to operate without adding staff. In these cases, program management has been shared among one or more of the existing staff. Larger facilities may require dedicated staff to oversee

TABLE 2.1 Program Management Categories and Approaches

Category	Element	Low Effort/Cost	High Effort/Cost	Notes
Program Staffing	Staffing changes	No changes (distribute duties to existing staff)	Hire additional staff	Quantify the number of hours required to manage program. Include administrative and field work.
Monitoring and Compliance	Sampling waste	Once per hauler at time of hauler permitting. Can request analytical from generator/hauler rather than have WRRF collect and analyze waste.	Online pH measurements. Sample every load, analyze samples routinely or on each load. Staff oversight at receiving station, including visual inspection of waste.	Periodic, if not routine, sampling is recommended to verify waste compliance with hauler permits. Confirm there are no toxic or inhibitory compounds or contamination. Routine sampling for process performance and planning may be required. Periodic sample collection also provides a deterrent for hauler to bring material outside of permit conditions.
	Record keeping	Record customer deliveries and volumes for billing purposes.	Recording volumes delivered for billing purposes and process/tipping fee analysis.	Program management review of records can inform operation decisions and tipping fee evaluation.

Operations and Maintenance	Co-digestion waste receiving station operations	No staff at receiving station.	Automated receiving station or fully staffed.	Staffed operations may have limited receiving hours compared to a fully automated station, which has higher design costs.
	Receiving station maintenance	Weekly, monthly, and periodic maintenance required for equipment; change pump parts and replace odor control media.		Maintenance requirements are impacted by design decisions.
	Biogas conditioning and utilization maintenance	Weekly, monthly, and periodic maintenance required for gas conditioning and utilization.		Certain maintenance items can be contracted to third party (e.g., engines).
Traffic Control and Site Security	Receiving station site location	Site located within WRRF, adjacent to digesters.	Site located outside of WRRF fence line.	Locating receiving station site outside of WRRF improves security but may be more effort for design/monitoring.
	Site security	No security requirement.	Security check-in for each hauler.	Haulers can be given access key cards coded for certain hours (medium effort).
	Traffic control	Add signage for truck route.	Develop new roadways for co-digestion traffic.	Consider impacts to WRRF staff safety if traffic volume increases. Consider conflicts with other traffic (chemical deliveries, biosolids, etc.).

program administration and for monitoring in the field. A program manager may be considered if the amount of waste and number of truck haulers are substantial. Administrative roles such as billing, approving/renewing waste hauler permits, and adjusting tipping fees are important to maintain a sustainable program. Staff will need to compile and analyze records from co-digestion waste deliveries to manage the trajectory of the program and evaluate whether the program will impact WRRF capacity. Field staff should be considered if regular sampling will be conducted. Field staff can also monitor and assist with hauler offloading, maintain receiving station equipment, and clean up spills.

2.2 Monitoring and Compliance

Monitoring can be periodic or continuous, depending on WRRF resources and staffing. Compliance is closely aligned with monitoring. Higher monitoring efforts increase the likelihood that the WRRF will observe noncompliant waste, but should decrease the number of bad actors in the long term. Monitoring protects the WRRF from process upsets and the community from adverse environmental impacts.

2.2.1 Sampling

Collecting and analyzing samples provides the ability to monitor the program and generates data for the program manager to estimate treatment costs. Sampling can be scheduled more frequently when a new waste type or industry is permitted to verify compliance and check variability of the waste; alternatively, samples can be collected and analyzed for every load. An intermediate solution could be sample collection and storage (e.g., a few days at 4 °C), and only analyzing samples if there is a process upset. The appropriate sampling schedule will depend on the size and history of the co-digestion program and the availability of staff/resources. There are three locations that a WRRF may typically sample co-digestion waste:

1. **Waste source**—samples may be provided by the industry generating the waste or collected by the WRRF at the generator location.
2. **Truck hauler**—samples may be provided by the waste hauler or taken by WRRF staff at the time of discharge. Samples could also be labeled and left by the hauler.
3. **Receiving station digester feed**—this sampling location will only be available for sampling if there is a co-digestion blending or holding tank at the receiving station. Sampling downstream of homogenization co-digestion waste streams provides a more consistent sample that is

better suited for evaluating the overall program impact on facility performance and capacity but will not provide data on individual loads.

All three sampling locations allow the WRRF to analyze parameters that can impact process performance (e.g., total and volatile solids, chemical oxygen demand, nitrogen, and phosphorus) and monitor toxic materials, such as metals (Water Environment Federation, 2024). A pH measurement is best collected at the waste generator or from the truck hauler to confirm an acceptable value.

2.2.2 Waste Hauler Compliance

Waste haulers are required to comply with all regulatory and WRRF permit requirements. The WRRF may elect to impose more stringent requirements than federal or state requirements based on the needs of the WRRF. Depending on issues that may be identified during monitoring and sampling activities, compliance checks may be as infrequent as required by regulatory requirements or aligned with the permit renewal (e.g., annually).

2.2.3 Recordkeeping

Proper recordkeeping is important to document compliance issues, field observations, and sampling results. Diligent input of all information on each waste discharged at the WRRF provides an accurate log of activities/discharges that may be necessary to collect revenue (i.e., billing) or troubleshoot upsets. Accurate records provide program managers with the ability to plan for future changes in the co-digestion program based on past performance. Maintaining accurate and organized records allows the WRRF to perform their own analyses or outsource analysis to an independent party (i.e., a consultant).

2.3 Operations and Maintenance

Operations and maintenance are critical for co-digestion program success. High-strength, organic wastes increase biogas production significantly, which can make existing biogas handling and treatment systems more challenging to operate. These wastes may also impact digester mixing or cause foaming. Impacts are described in more detail, including operating strategies, in subsequent chapters. Traditional operator licenses should be sufficient for operation of receiving stations, and staff monitoring hauler offloading may not require an operator license if there is a receiving station. The digester feed pumps are the tie-in to WRRF operations.

Maintenance impacts from co-digestion should also not be taken lightly. A successful co-digestion program will require additional effort from maintenance staff. Receiving stations require daily, weekly, and periodic maintenance to clean out pretreatment (e.g., rock traps), maintain instrument and equipment cleaning and greasing schedules, and change out damaged equipment (e.g., replacement of pump components). Wastes are typically acidic and may contain high levels of abrasive contaminants, such as rocks, concrete, and grit. Abrasive materials not only wear down receiving station equipment, but can require more frequent maintenance of solids treatment processes downstream (e.g., digester cleaning, dewatering equipment, solids treatment pumping equipment). Co-digestion may increase nutrient concentrations and exacerbate phosphorus precipitation within pipes and equipment. Cleaning and replacement of assets impacted by struvite is a nuisance and increases operating costs. Nuisance odors typically require an odor control system, which may require frequent media changeout.

2.4 Traffic Control, Site Safety, and Security

Traffic control may require staffing if waste haulers do not follow posted signs. Drivers that exceed posted speed limits present a safety hazard for existing staff, especially for co-digestion programs of significant size. Below are three site access approaches that can be implemented:

- **Open access**—open access can be provided to waste haulers during normal business hours. This requires the least amount of effort from staff, but involves the highest risk. WRRFs that typically leave gates open during normal business may elect to operate in this manner.
- **Automated check-in**—use of card readers allows waste haulers to check in and open gates with automated systems. These systems can be programmed to allow access during specific days or hours within the week.
- **Physical oversight**—a dedicated check-in location allows the WRRF to staff security guards for the most secure site access option. The check-in location can be paired with sampling and recordkeeping, using the staff for both security and monitoring.
- **Remote oversight**—cameras can be set up to provide real-time monitoring of deliveries or to investigate incidents after the fact.

2.5 Tipping Fees

Tipping fees provide the WRRF with revenue to offset operating and maintenance (O&M) costs associated with co-digestion waste and residuals so revenue exceeds cost to treat. Tipping fees should consider both waste volume and characteristics. Tipping fees based on volume can be structured in different ways:

- **Tank capacity**—fixed based on tank capacity for each hauler. This structure is the simplest because it does not require the WRRF to measure the delivered volume.
- **Delivered volume**—variable, based on measured volume of waste received by the WRRF. The volume of waste received would need to be quantified via flowmeter, differential level at the receiving station, through weight tickets, or otherwise. Potential errors in volumetric measurements or calculations and additional burden of measurements may make this structure less attractive.
- **Tiered structure**—based on total volume delivered per period of time. For example, charging a higher rate for less than 37 850 L/d (10 000 gal/d) of waste and a lower rate for over 37 850 L/d (10 000 gal/d) of waste. This structure may be used to incentivize higher quantities of certain waste types (e.g., fats, oils, and grease) if there is regional competition for co-digestion wastes, or incentivize deliveries at beneficial times (e.g., weekends).

Differences in waste characteristics can lead to variable impacts on O&M costs; different levels of inert solids lead to more grit and biosolids production, and different organic loading results in greater biogas production. Concentration-based tipping fees account for constituents in different types of wastes:

- **Concentration**—based on a constituent in the waste. This structure may be used to more accurately align revenue and cost to treat based on treatment costs for residuals and/or nutrients. Alternatively, this structure can be implemented to incentivize wastes with higher biogas potential per unit volume to decrease the impact on digester hydraulic retention time. Calculating billing based on loading is more complicated and time-consuming for the WRRF staff. A tiered structure can also be applied to concentration-based fees.

Tipping fees are typically set before co-digestion program startup. The WRRF can communicate tipping fees with industries and haulers via direct outreach, the permit application, or the agency's website. Tipping fees will need to be evaluated regularly to keep up within escalating O&M costs. Tipping fees may also need to be adjusted to account for market competition. In some cases, geographical tipping fee structures can be implemented where a lower tipping fee applies to waste collected within the service area. Annual tipping fee evaluation and adjustments allow WRRFs to align changes with their fiscal year. Governing body approval of tipping fees may be required. Providing advance communication of changes with industries and haulers allows them to adjust their business plan based on changes to their operating costs.

2.6 Enforcement

The WRRF should have established procedures to take enforcement action, similar to the procedures the WRRF uses to administer its pretreatment program. Violations could include a waste hauler delivering non-compliant waste. Actions may include imposing fees as a penalty, suspending a permit, or—if severe and repeated—revoking a permit.

3.0 BUSINESS CASE EVALUATION

A profitable or sustainable business case evaluation is often a prerequisite before starting a co-digestion program. Receiving and processing additional waste at a WRRF produces additional biogas, but there are also increased costs associated with staffing, treatment, and maintenance. Table 2.2 presents capital, operating, and life-cycle cost components that should be considered in a business case evaluation. A business case evaluation compares one or more co-digestion scenarios against a "do nothing" scenario, where co-digestion is not implemented. If life-cycle costs are lower for co-digestion than the status quo, then this represents a sustainable business case evaluation. In some cases, co-digestion programs can provide a profit to WRRFs within a 10- or 20-year period.

3.1 Capital Costs

Capital costs can make up the majority of life-cycle costs for co-digestion program, especially if a shorter return period is used for the evaluation. The primary and most obvious cost is the receiving station, which is a best practice for successful co-digestion operations. If capacity improvements are required to implement co-digestion, then the business case evaluation may show that a program is cost-prohibitive; co-digestion programs are typically implemented to take advantage of existing excess capacity. Germany's co-digestion programs demonstrate this phenomenon (Rauch-Williams & Schaum, 2018). Depending on the size of the co-digestion program, additional improvements to site civil or administrative buildings may be required. If significant truck traffic will be added to the WRRF, then

TABLE 2.2 Business Case Evaluation Components

Capital Cost Components	Operating Cost/ Benefit Components	Life-cycle Cost Components
Receiving Station Capacity Improvements • Digestion • Dewatering • Gas conditioning • Gas utilization • Gas flare/waste • Nutrient removal Site Civil • Roadway improvements • Fencing/Security Administration Buildings • Additional space for program staff office/ additional laboratory work	Staffing/labor to issue permits; oversee program; enforcement; monitoring Maintenance staff to manage increased wear on equipment Rehabilitation and replacement of receiving stations; digestion; dewatering; and biosolids management Digestion/Solids Treatment • Heating • Mixing • Pumping • Polymer Biosolids Hauling/End-use Tipping/permitting fees (revenue) • Energy production (revenue) avoided cost and value of electricity sold • Environmental attributes	Net present value duration (years) Discount rate Escalation rates • Electricity • Chemicals • Tipping fee revenue • Natural gas

improving roads to accommodate heavier trucks and potential spills may be prudent. If additional staff is added, facilities to accommodate workspaces or hauler check-in buildings may need to be added.

3.2 Operating and Maintenance Costs

A close look at realistic O&M costs is necessary for more accurate life-cycle costs analysis. The primary and most obvious O&M costs are associated with oversight of the receiving station and increased maintenance costs (staff time and materials). Changes in operating costs are due to residuals from co-digestion wastes in digesters (i.e., grit accumulation) and digestate (i.e., additional biosolids, increased nutrients). Increased nutrients in effluent discharge can impact liquid treatment operation costs (e.g., additional aeration or supplemental carbon addition), and higher nutrient concentrations or loads in biosolids can impact land application practices and increase biosolids management costs (Jones et al., 2019). Some facilities have documented negligible changes in solids treatment chemical costs (e.g., ferric and polymer), but changes in chemical requirements for either biogas conditioning, dewatering, or struvite mitigation should be considered in the economic evaluation (Appleton & Rauch-Williams, 2017).

The co-digestion program offsets O&M costs by providing cost benefits if biogas is beneficially used to generate revenue (e.g., renewable identification numbers or lower energy costs (e.g., biogas use in engines or turbines for combined heat and power generation [cogeneration]). Tipping fees, which are discussed in greater detail below, also generate revenue. Permitting revenue can also be collected.

3.3 Life-cycle Costs and Triple Bottom Line

A life-cycle cost analysis will help determine whether co-digestion is economically feasible to implement. Life-cycle costs can be evaluated based on net present value, which summarize all capital costs and annual costs/benefits. The net present value should demonstrate a net positive revenue (i.e., no cost) to justify implementing a co-digestion program. Co-digestion programs have demonstrated payback typically within a 10- to 20-year period, but are dependent on location-specific conditions and should consider sensitivity (Higgins & Rajagopalan, 2017).

Triple bottom line expands on life-cycle costs and incorporates impact to society and the environment. This approach weighs non-economic impacts and may be important to purpose-driven entities, such as WRRF agencies. The TBL approach can be used to quantify impacts on the WRRF staff and the community (including industries that will participate in the co-digestion program). This approach also allows the WRRF to determine impacts to the climate, receiving-water body, and land. Additional resources for learning net present value and TBL are provided at the end of this chapter.

3.4 Tipping Fee Evaluation

Tipping fees may be necessary for an economically viable co-digestion program. Tipping fees should be evaluated such that costs to treat waste residuals are covered by the tipping fees, and even higher fees may be necessary to offset capital costs. The tipping fee for each waste type may need to be adjusted during the business case evaluation based on the value needed for a break-even life-cycle cost. The tipping fee upper limit will be dictated by both market competition and current industry/hauler disposal costs.

3.4 Risks and Considerations

Conducting a sensitivity analysis is critical. The sensitivity analysis should be set up to thoroughly evaluate all program risks, including

- **Capital costs**—capital cost estimates should have the proper contingency applied during program planning, and estimates should be escalated to account for escalation. Sensitivity for inflation or a noncompetitive bidding environment should be applied.
- **Operating costs**—although some facilities report minimal impacts to chemical addition and biosolids production, sensitivity should account for realistic worst-case scenarios for changes in biogas quality, dewaterability, biosolids quantity, and nutrient recycle. Later chapters discuss operational impacts from co-digestion and can be referenced to determine bounds for the sensitivity analysis. Sensitivity can also include additional full-time staff if needed for operation of the receiving station or program management.
- **Maintenance costs**—accounting for more frequent repair and replacement (R&R) of wearable equipment components, such as pump rotors and impellers, should be included in the sensitivity analysis. Contamination may pass through receiving stations and impact the existing solids treatment process. Previous experience from facilities with co-digestion provide a typical range of R&R requirements (Appleton & Rauch-Williams, 2017), and a higher frequency should be considered in the sensitivity analysis.
- **Biogas production and utilization**—biogas production will increase with co-digestion, but a sensitivity analysis should consider whether variable production will impact utilization. A higher percentage of wasted biogas can be assumed for the sensitivity analysis. This will help identify minimum operating goals for the WRRF based on the business case evaluation. The cost of electricity also has significant impact on the evaluation. The sensitivity analysis should consider higher prices of electricity, and determine the impacts. If program implementation is not feasible currently, this parameter (energy unit cost) may be tracked to identify the trigger for co-digestion program reevaluation.

- **Volume received**—consider the impact on life-cycle costs if lower volumes of waste are received than expected. If multiple waste types are expected, then the sensitivity analysis can determine which waste has the biggest impact on life-cycle costs. This will help identify risk which industries and haulers are most important to retain.
- **Tipping fees**—the tipping fee can be adjusted based on O&M costs or market competition. A lower tipping fee should be evaluated to account for future market competition if there are more local co-digestion programs. The sensitivity analysis should also consider life-cycle costs if no tipping fee is charged, to account for a worst case scenario.
- **Regulatory changes**—consider the impact on treatment and disposal costs of co-digestion residuals if regulations change and require additional dewatering return treatment or alternative biosolids disposal.

Aside from economic impacts, the WRRF should consider culture and morale impacts from a co-digestion program on the agency and staff. Support from all stakeholders is necessary to facilitate a successful co-digestion program. Expanding existing staff responsibilities will likely be necessary to accommodate additional management, administration, operating, maintenance, sampling, and analytical requirements. Community engagement may also be necessary if increased traffic is expected to impact the surrounding community. Program acceptance within the agency and local community should not be underestimated and is important for long-term sustainability.

4.0 REFERENCES

Appleton, R. A., & Rauch-Williams, T. (2017). *Co-digestion of organic waste addressing operational side effects*. Water Environment & Reuse Foundation.

Global Methane Initiative. (2014). *A global perspective of anaerobic digestion policies and incentives*. Global Methane Initiative Agriculture Subcommittee.

Higgins, M., & Rajagopalan, G. (2017). *Understanding impacts of co-digestion: Digester chemistry, gas production, dewaterability, solids production, cake quality, and economics*. Water Environment & Research Foundation.

Jones, C. A., Coker, C., Kirk, K., & Reynolds, L. (2019). *Food waste co-digestion at water resources recovery facilities: Business case analysis*. Water Research Foundation.

Rauch-Williams, T., & Schaum, C. (2018). *Co-digestion experience in Central Europe and case study analysis*. Water Research Foundation.

U.S. Environmental Protection Agency. (1999, September). *Guidance manual for the control of wastes hauled to publicly owned treatment works*. https://www.epa.gov/system/files/documents/2021-07/hwfinal.pdf

Water Environment Federation. (2024). *Anaerobic co-digestion: Feedstock sampling and characterization fact sheet*. Water Science and Engineering Center.

5.0 FURTHER READING AND RESOURCES

Along with references listed above, the reader is directed to the following resources for additional information that may be useful when planning a co-digestion program, including creation of hauled waste permits and conducting a business case evaluation. Subsequent chapters provide additional details necessary to determine operational impacts and design guidelines.

Center for EcoTechnology. (n.d.). *Northeast anaerobic digester accelerator – Food waste digestion insights*. Accessed May 2024. https://wastedfood.cetonline.org/wp-content/uploads/2022/09/WFS-Insights-on-AD-Tech.pdf

Co-digestion economic analysis tool information checklist. (n.d.). Accessed July 2024. https://archive.epa.gov/region9/organics/web/pdf/checklist_draft.pdf

Gallo, A. (2014). A refresher on net present value. *Harvard Business Review*. Accessed July 2024. https://hbr.org/2014/11/a-refresher-on-net-present-value

Miller, K. (2020). The triple bottom line: What it is & why it's important. *Harvard Business School Online. Business Insights Blog*. Accessed July 2024. https://online.hbs.edu/blog/post/what-is-the-triple-bottom-line

State of Oregon Department of Environmental Quality. (n.d.). *Hauled waste plan review checklist*. Accessed May 2024. https://www.oregon.gov/deq/FilterDocs/indpret-haulerwastechecklist.pdf

Chapter 3

Feedstocks Evaluation and Selection

Bhargavi Subramanian, PhD; Yamrot A. Amha, PhD, PE; Vanessa Borkowski, PE; Jayesh Carthal; Joe Cleary; Felipe Gutierrez, PhD, PE; Ed Horai; Erik Larson, PE; Ding-Hsiang (Shawn) Lin; Michaela McKenzie, PE; & Leila Tootchi, P.Eng.

1.0 TYPICAL FEEDSTOCKS USED IN CO-DIGESTION

Over the last two decades, utilities have increasingly sought to improve their sustainability and move towards energy neutrality. One approach to becoming energy self-sufficient is to increase biogas production by seeking alternative feedstocks such as farm wastes, food wastes (FWs), and fats, oils, and grease (FOG), and co-digesting them with the sludges already produced at the facilities. Such feedstocks are classified as high strength wastes (HSWs) by virtue of their high chemical oxygen demand (COD) and solids loading. Such feedstocks are also chemically very different from sewage sludge. Industrial wastes and wastes from the processing of foods that are high in sugars or lipids are all classified as HSW feedstocks.

There is an abundance of literature published on laboratory-, pilot-, and full-scale studies related to co-digestion. The focus of this chapter is to provide an overview of the various feedstocks and their characteristics and provide guidance in feedstock selection.

1.1 Introduction of Feedstocks

Organic feedstocks that are typically added to the digesters for co-digestion are more readily biodegradable and have a volatile solids (VS) destruction rate of 80%–90% (Higgins et al., 2017). This is higher than conventional wastewater sludges which are typically 45%–55% (Coyne et al., 2017), and the higher VS destruction rate results in increased biogas production. Feedstock addition for co-digestion can either improve or negatively impact digestion performance depending on its composition, loading rates, digester performance, and synergy between the sludge and the feedstock. An in-depth discussion of the effects of co-digestion is described in Chapter 4.

1.1.1 Fats, Oils, and Grease and Grease Trap Waste

Fats, oils, and grease and grease trap waste (GTW) are typical feedstocks that may be fed to the digesters along with sludge for co-digestion. Yellow or brown grease that is collected from restaurants and other food facilities is often the most calorically dense making it a desirable feedstock. Fats, oils, and grease that are often collected via grease traps that intercept sanitary sewage before disposal within the collection system are termed GTW. This originates from restaurants, cafeterias, and other commercial or industrial customers and is typically lower in strength than the FOG due to dilution. Grease trap wastes are also generally referred to as FOG. Fats, oils, and grease are typically received directly via FOG haulers and screened and diluted before being fed to the digesters. Detailed description of feedstock receiving is described in Chapter 5. Grease trap waste from restaurants were one of the first feedstock targets for co-digestion. This grease contained fats, oils, and grease with a high energy value as measured by COD and BOD. It has one of the highest biomethane yields of all the feedstock being used as will be discussed later. Grease trap wastes/fats, oils, and grease have been used in many co-digestion projects for 10-plus years now in Hershey, Pennsylvania; West Hermitage, Pennsylvania; Los Angeles, California; and Ridgewood, New Jersey, to name a few, and such case studies are discussed in Chapter 9.

Some of the key design issues include pretreatment to remove debris such as cooking utensils, napkins, and plastics that are harmful if they get in the digesters. The pretreatment typically includes screening before equalization and storage tanks which as used to control the feeding rate to the digesters. Another design issue is that in colder climates, it is necessary to heat the equalization and storage tanks before the digestion process. This improves viscosity and feed flow quality to the digester.

A third issue is the feeding rate to the digester. Feeding too much grease versus primary and secondary sludge can create problems with the biomass. This recommended ratio should be evaluated based on relevant literature in combination with laboratory bench testing

1.1.3 Animal and Related Wastes (Poultry, Meat Processing, Dairy)

Animal wastes include solids and liquids from various animal processing operations, including slaughterhouses and dairy farms. The byproducts from these operations can include meat processing waste (MPW), blood, and dairy waste. These particular feedstocks are of interest due to reports from sources such as the American Biogas Council and Canadian Biogas Association that indicate the potential for thousands of anaerobic digestion facilities to come to fruition based on these available feedstocks.

Meat processing waste is a lipid-rich source that may facilitate greater biogas production. In one study that was completed for three different proportioned blends of waste activated sludge (WAS) and MPW, the blend of 50% WAS and 50% MPW, by mass, produced significantly increased the biogas volume per g VS when compared against the control (Cao et al., 2024). Furthermore, this study also measured the microbial populations among the WAS and MPW blends, and demonstrated that the blend of 50% WAS and 50% MPW resulted in a relatively higher representation of methanogens. For applications that co-digest MPW, the study reinforces that pretreatment for the feedstock, meaning disintegrating it into smaller pieces, is an important process to allow the microbial community within the digester better access to lipids.

Another type of valuable animal waste is blood. In particular, the livestock slaughtering process generates large amounts of blood as a byproduct. Blood has a high COD and nitrogen content and is therefore preferred to incorporate into anaerobic co-digestion rather than the main liquid treatment process. Direct discharge of untreated blood in the wastewater liquid process increases the organic pollution load on wastewater by up to 35%–50% (Nazifa et al., 2021). This type of organic pollution load is harmful to the environment if it is discharged without prior treatment. Furthermore, most industrial and commercial entities are regulated at the local level, and the co-digestion facility may negotiate the feedstock quality within the terms of the contract with the feedstock generator (U.S. Environmental Protection Agency [U.S. EPA], 2012). When processing livestock such as chicken, only 75% of the bird's weight can be used for edible meat while the remaining 25% is discarded as byproducts (feathers, blood, intestine residues). This applies to other forms of animals processing on a daily basis. However, blood is valuable as it is mainly composed of protein and a small percentage of lipids. Its highly degradable protein content contributes to a high methane yield (Nazifa et al., 2021). According to an article by Afazeli et al., from 54×106 m^3 of the annual biogas yield potential from waste at animal processing facilities, up to 31% of the biogas may be produced from blood (Afazeli et al., 2014). Generally, slaughterhouses only recover on average 3%–4%, 3%–4%, and 3.5%–4% of blood from pigs, cattle, and lamb, respectively (Nazifa et al., 2021).

Dairy waste such as waste milk, whey, and lactose have also been studied as potential feedstocks for co-digestion. One study conducted at a facility in Germany co-digested dairy waste at the water resource recovery facility (WRRF), and the specific methane yield for the dairy waste products was 20% to 60% greater than that of primary sludge from the facility (Semberg et al., 2019). Paired with higher organic loading rates for the feedstocks, the reported overall biogas production was estimated to increase by 300% ± 50%. With much of the dairy waste being provided in liquid form, pretreatment is not as much of a concern for this feedstock. However, the disadvantages associated with co-digesting dairy waste include increased nitrogen backload and reduced dewaterability (Semberg et al., 2019). The nitrogen backload may be managed with sidestream treatment. Furthermore, the dewaterability may be less of a concern if food waste is not a co-substrate with dairy waste.

1.1.3 Agricultural Wastes

Agricultural wastes fall under two types: primary and secondary (Adnane, 2024). Primary residues are also called field-based residues and are generated as by-products of crop post-harvesting. This includes cereal straw, maize stalks, stubble, leaves, and branches. Secondary or process-based residues are obtained during industrial crop processing. These include nut shells, almond husks, olive stones, and coffee pulp (Adnane et al., 2024). Other residues are forest residues and wood wastes as well as vegetable and fruit residues.

In general, the potential of crop residues is directly proportional to total crop production. However, this potential varies from crop to crop and from one region to another due to numerous factors including the type of crop grown, soil type, local meteorological conditions, and harvesting techniques (Adnane et al., 2024). However, not all estimated agricultural residues are collectible and useful for bioenergy production. Only the surplus residue potential can be harnessed due to competing applications such as livestock fodder and bedding, household fuel for cooking and heating, roof thatching, as well as for ecosystem services like soil mulching, conservation of soil fertility, and erosion control (Adnane et al., 2024).

Specifically, crop residues are differentiated from other residues due to their high organic loading, a high carbon-to-nitrogen ratio, but a low buffering ability.

The type of substrate and its biochemical composition are also crucial to consider when evaluating digester performance, biogas yield and composition, and the overall effectiveness of the anaerobic co-digestion process. While protein-rich substrates such as animal manure and substrates from dairy and meat processing industries yield greater methane potential, their microbial degradation often leads to ammonia buildup. This buildup can destabilize the fermentation and even lead to process failure. To address this issue, utilizing carbohydrate-rich substrates (lignocellulosic residues, food wastes, and vegetable wastes) is an efficient tactic to prevent such problems. Lipid-rich feedstocks, such as palm oil mill effluents and olive oil mill effluents, have a higher theoretical methane potential than carbohydrate-rich and protein-rich substrates (Adnane et al., 2024). However, the degradation of lipidic waste produces long-chain fatty acids and in high concentration, they can impede methanogenic metabolism, clog the mass transfer between anaerobes and the media, cause sludge washout, and lead to the formation of scum and foam. Therefore, the use of carbohydrate-rich substrates as co-substrates in the anaerobic digestion process can alleviate the inhibitory effect of long-chain fatty acids and improve process efficiency (Adnane et al., 2024).

The pretreatment of residues is crucial for enhancing substrate digestibility within the bioreactor and ultimately improving biogas production. Therefore, the pretreatment stage is necessary to optimize biogas yield through breaking down the recalcitrant structure of the feedstock and making polysaccharides more accessible to enzymes and microorganisms (Adnane et al., 2024). Pretreatment technologies include physical, mechanical, thermal, chemical, and more. Ultimately, the pretreatment's effectiveness is influenced by factors such as the physio-chemical properties, economic and energy costs, environmental impacts, and methane improvements achieved. With physical pretreatment, there are no additional compounds added and it is a well-known method used to improve biogas yield. This method increases the substrate's surface area and improves contact between the substrate and microorganisms (Atelge et al., 2020). Mechanical pretreatment is known to increase bulk density, porosity, and surface area of a substrate. The waste structure characteristics are altered through this process, such as structure of lignocellulosic materials. The main goal of this method is to decrease particle size, thereby increasing surface area for the lignocellulose substrate. Hammer mills and knife mills are common for this type of pretreatment. Another well-known and established method is thermal pretreatment, which is dependent on time, temperature, and pressure. This method is arguably the most established and through this process, the substrate structure is changed from insoluble to soluble through heat application. As a result, the substrate's biodegradability increases (Atelge et al., 2020).

Co-digestion of WAS and agricultural straw waste was conducted and biochemical methane potential (BMP) of co-digestion of waste activated sludge with three agricultural wastes: wheat straw, rice straw, and soybean straw were carried out. Their results infer that the co-digestion system of WAS with wheat straw or rice straw or soybean straw yielded higher cumulative methane production of 223, 218, and 197 mL/g VS, which produced 2.57, 2.52, and 2.27 times higher methane yield compared to the mono-digestion of WAS respectively (Potdukhe et al., 2021). The results are a promising step toward determining best practices for agricultural waste use in co-digestion.

1.1.4 Food Wastes

Food wastes are a widely used feedstock for co-digestion. Food wastes constitute the biggest portion of source separated organic (SSO) feedstock. Pre-consumer feedstock originates from food sources before supply to and use by consumers and wasted due to overproduction, spoilage, sub-standard quality, or expiration. It includes sub-quality food sorted out by grocery stores, items that have exceeded their shelf life, and waste from food processing facilities and warehouses. Post-consumer feedstock originates from food after consumer use. It includes residential food waste as well as food waste from restaurants, canteens, cafeterias, and so forth. In the United States, the SSO feedstock received by the WRRFs may include a mix of both pre- and post-consumer food waste depending on the suppliers and the geographical location. Most WRRFs prefer to procure pre-processed food waste that has been cleaned and processed by waste processors. One such cleaned food waste source engineered BioSlurry. Several WRRFs in the United States receive an engineered BioSlurry product that has consistent quality (Rauch-Williams et al., 2023). Bioslurry may be made from pre- or post-consumer food source depending on availability of the raw feed material. Table 3.1 lists the available feed characteristics of various food waste feedstocks. From the table, it can be seen how varied the food waste characteristics may be depending on the region, processing method, and sources of the raw waste.

1.1.5 Brewery and Winery Wastes

There are three main solid waste streams in the brewery process: brewery spent grain (BSG), hot trub, and residual yeast. Brewery spent grain is produced in large quantities in the brewery industry, with estimates of around 30–44 lbs of waste per 26 gal of produced beer, representing 85% of the byproducts produced by any brewery (Mallen & Najdanovic, 2018; Szaja et al., 2020). Hot trub is generated through precipitation from wort during wort boiling and hops addition process at a rate of 0.4–0.8 lbs per 26 L of produced beer, while residual yeast is excess yeast produced during the fermentation process (Mathias et al., 2015). These waste streams are characterized by high moisture content and high organic matter, therefore, are widely recycled in agriculture, either as animal feed or as fertilizer.

There is a growing interest in using BSG as substrate in the co-digestion process due to the high-volume generation of this waste stream. A study (Mathias et al., 2015) that assessed the composition of total fraction of BSG in dry weight basis estimated 3.8% ash, 52.3% total organic carbon, 4.3% total nitrogen, and 26.9% total phosphorus (TP). In addition, the study estimated the moisture content of BSG to be 82.6% and carbon to

TABLE 3.1 Food Waste Composition Reported by Various Authors

FW Type and Origin	TS [%]	VS [%]	Proteins [%]	Lipids [%]	Carb [%]	tCOD (mg/L)	C [%]	N [%]	Reference
Food waste		29.3	26.6	35	32.5		48.4	3.8	Morales-Polo et al., 2018
Food waste	18.1	17.1		23.3	61.9				
Food waste	23.1	21.0					56.1	2.3	
Food waste for anaerobic digestion	30.9	26.35					46.8	3.54	
Kitchen waste	24	23.2	15	23.9	55.2		54	2.4	
Food waste	23.2	21.7	2.9	6.5	13.7				
Organic fraction from municipal biowaste	20	18	30	4.4	10.7				
Food waste	16.7	15.3	2.4	1.4					
WM CORe® Bioslurry	12.9	89.4 (% of TS)				222 482			Rauch-Williams et al., 2023
U.S. cafeteria food waste	23.8	90.9 (% of TS)				197 000			
Pre-processed food waste from Waste Management, California						200 000			
Food waste						139 667	53	5	Rajagopalan et al., 2022
Brewery		97 (% of TS)				88 100	46	2	
FOG		93 (% of TS)				48 700	63	2	
Creamery waste		94.6 (% of TS)				305 000	67.4	1.6	
Poultry blood waste		92.8 (% of TS)				210 400	50	14	
Blended whey		87				76 000	38	1	
Poultry dissolved air flotation waste		96.5				287 600	65	4.9	

nitrogen (C/N) ratio at 12.1. Mono-digestion of BSG was found to be ineffective due to the limited biodegradability of lignin and demand for long hydraulic retention times (HRT), but digestion of BSG is substantially improved by using co-digestion with substrates such as cattle dung, ping manure, and sewage sludge (Szaja et al., 2020). As a co-substrate in a co-digestion process, BSG provides significant biogas potential, beneficial C/N ratio, and high presence of mineral salts, vitamins, and amino acids, which may be present in low levels in sewage sludge (Szaja et al., 2020). Thus, several studies have reported improved biogas production when BSG is used as substrate in co-digestion process (Goberna et al., 2013; Polastri et al., 2024). The specific methane yield of BSG can range from 0.064 to 0.550 Nm^3/kg VS, considering different factors such as type of pretreatment applied, reactor configuration, temperature, and composition of the substrate in terms of cellulose, hemicellulose and lignin content (Bachmann et al., 2022).

Major waste streams produced from winery include marcs (or pomace) from pressing, wine lees from settling, stalks, from destemming, and winery wastewater. With 2204 lbs of processed grapes to produce 198 gal of wine, the specific waste flows are estimated to be 286 lbs of marc, 132 lbs of wine lees, 66 lbs of stalks and 436 gal of wastewater (Chiappero et al., 2023). Generally, winery waste streams are characterized by low pH, high organic content, low concentration of nitrogen, and potential for presence of compounds such polyphenols that can potentially be toxic to the anaerobic digestion process. Thus, co-digestion of winery waste has the potential to mitigate some of the challenges of mono-digestion of winery waste by increasing stability and providing missing nutrients, and diluting recalcitrant compounds like polyphenols and copper (Da Ros et al., 2014; Ripoll et al., 2020).

Grape marc is the waste remaining after grape juice extraction and consists of residue of grape skin, pulp, and seeds. This waste stream is characterized by high content for organic compounds, carbohydrate, and phenolic compounds, with low concentrations of nitrogen and phosphorus. Grape marc contains high concentrations of lignin (16.8%–24.2%) and cellulose (27%–37%) and moisture content of about 65% (Ahmad et al., 2020; Fabbri et al., 2015). Co-digestion has the potential to improve digestion of grape marc waste. A study (Kassongo et al., 2020) that evaluated co-digestion of grape marc and cheese whey reported maximum methane yield of 0.4 cu ft methane/lb of VS.

Wine lees are generated due to the separation of solids from wine and consist of fine fruit pulps, tartrate salts, spent yeast, bacteria, and debris, and typically 13.2 lbs of lees are produced per 26.4 gal of produced wine (Ros et al., 2014). The methane potential of wine lees can range between

6.4 to 9.6 cu ft/lb COD removed, with methane content of 60% (Da Ros et al., 2014). Treatment of grape marc and wine lees for a medium sized winery (1000 tones grapes per season) with compact anaerobic digesters was shown to be economical viable, by resulting in biogas production of 8 cu ft/lb COD at methane content of 72% and payback period of around 6 to 7 years (Eleutheria et al., 2016).

1.2 Main Components of the Feedstocks

1.2.1 Proteins and Amino Acids

Protein-rich streams, commonly found in dairy and slaughterhouse wastewater, have been considered potential feedstock for biogas production. Previous studies have shown that protein-fed reactors can operate stably and produce biogas; as such, the protein-rich stream can be used for bioenergy and ammonia recovery (Deng et al., 2023).

Anaerobic protein degradation can be summarized into three steps: hydrolysis of proteins to amino acids, deamination (or acidification) of amino acids into ammonium and volatile fatty acids (VFAs), and methane (CH_4) production from VFAs (i.e., methanogenesis) (McInerney, 1988).

Based on prior research, the presence of proteins can cause some problems with anaerobic digestion, such as foaming, incomplete degradation of organic nitrogenous compounds (Bareha et al., 2018), an increase in total ammonia nitrogen concentration leading to the inhibition of methanogenesis (Jiang et al., 2019), and deterioration of morphological sludge properties (Liu et al., 2019). Another issue is that proteins are degraded into ammonia, resulting in an effluent stream with a high nitrogen content. The stream usually has a flow rate of 1%–3% of the entire WRRF, but contributes to 15%–30% of the total nitrogen loading (Muszyński-Huhajło et al., 2021).

1.2.2 Lipids and Fats

The addition of lipids and fats-rich waste from several food and other processing industries (e.g., dairy, slaughterhouses, edible oil production, fish canning factories, bioethanol and diesel production, and wool scouring) to the anaerobic digestion of common solid wastes (e.g., sewage sludge or agricultural residues) has been shown to benefit biogas production rates. Although the benefit of increased biogas production potential sounds alluring, there are some problems associated with the co-digestion of lipids and fats. For example, microbial inhibition, and adsorption to the sludge lead to sludge flotation and biomass washout (Holohan et al., 2022). Therefore, it is critical to understand the properties of lipids and fats and how they interact with the anaerobic digestion process to fully utilize their potential.

Lipids and fats are a diverse range of compounds; however, they can be divided into four groups that are commonly found in wastewater: triacylglycerols, including long chain fatty acids (LCFAs), glycolipids, phospholipids, and cholesterol (Holohan et al., 2022). Among the four categories, triacylglycerols and LCFAs, commonly called fats and oils, are the most abundant (Holohan et al., 2022).

Under anaerobic digestion, fats and oils (e.g., triacylglycerols) are first degraded to glycerol and free LCFAs through hydrolysis. Glycerol is then converted to acetate by acidogenesis microbes. The released LCFAs, through a series of β-oxidation reactions, are converted to acetate (and propionate for odd-numbered LCFAs) and hydrogen (Cirne et al., 2007). As the last step, acetate and hydrogen are converted to methane through methanogenic microorganisms.

Theoretically, lipids are an ideal feedstock for anaerobic digestion as they have been shown to produce more biogas per weight of substrate (1.4 liters of biogas per gram of lipids) than other substrates (0.9 liters of biogas per gram of proteins and 0.8 liters of biogas per gram of carbohydrates) (Holohan et al., 2022). However, researchers have found that LCFAs can interfere with microbial communities and, therefore, with anaerobic digestion through two potential mechanisms: bacterial toxicity and temporary inhibition. Bacterial toxicity is hypothesized to stem from the long hydrophobic alkyl chain impacting the archaeal membrane, leading to irreversible cell decay. Temporary inhibition, on the other hand, hypothesizes that the LCFAs accumulate on the methanogenic microbes and prevent the mass transfer from the liquid to the microbial cells, but the cells' integrity and viability can be maintained. The methanogenic activity can be restored by eliminating the mass transfer limitations (Holohan et al., 2022).

To overcome the limitations of the anaerobic digestion of lipids and fats, commercially available bioreactors (e.g., bulk volume fermenter [ADI-BVF®], anaerobic flotation reactor [AFR], anaerobic membrane bioreactor [AnMBR], anaerobic sequencing batch reactor [AnSBR], and inverted anaerobic sludge blanket [IASB]) that use flocculent sludge have been developed over the last few decades. These innovative bioreactors have shown the possibility of treating lipid-rich wastewater anaerobically and high biogas production rates accompanied by saving the cost for lipids and fats pretreatment (e.g., screening, centrifugation, sedimentation, dissolved air flotation, flocculation, or precipitation). In addition, significant efforts have been made in the field of co-digestion with lipids and fats, mainly focusing on co-substrates ratios evaluations and correlating microbial community to process changes, to develop resilient practices to treat lipids and fats-rich wastewater and waste (Holohan et al., 2022).

1.2.3 Carbohydrates

In general, feedstock characterized by C/N ratio (discussed further in Section 1.3.1) of higher than 50, including wheat straws, corn stalks, rice, and algae, an organic fraction of municipal solid waste, could be used to supplement feedstock with lower C/N ratio to achieve nutrient balance and to avoid inhibition leading to system instability and reduced biogas production. Municipal wastewater sludge often has a low C/N ratio, this can lead to ammonia inhibition. Adding carbohydrate-rich waste can balance the C/N ratio, optimizing microbial activity and enhancing digestion stability. A combination of municipal sludge and organic fraction of municipal solid wastes has been proposed for creating a more balance C/N ratio for enhanced biogas production and methane yield in anaerobic co-digestion (Sosnowski et al., 2003). While a higher methane yield is attributed to lipids, carbohydrates have been associated with faster startup of gas production (de los Reyes et al., 2022). Additionally, the digesters that are started with carbohydrates develop microbial communities that are more robust to substrate changes compared to digesters that are started on protein based streams (de los Reyes et al., 2022).

Carbohydrate-rich streams have been co-digested with protein-rich streams to reduce inhibition of methanogenesis. Presence of protein could interfere with anaerobic digestion due to an increase in total ammoniacal nitrogen concentrations, which could in turn result in inhibition of methanogenesis (Jiang et al., 2019). Inhibition of ammonia/ammonium on methanogens, and the effectiveness of co-digestion with carbohydrate-rich streams, in combating that inhibition has been the focus of many studies (Mata-Alvarez et al., 2014; Rajagopal et al., 2013). In addition to a positive effect on methanogenesis, carbohydrates' co-digestion positively affects protein hydrolysis and negatively affects the deamination rate (Deng et al., 2023).

1.2.4 Fiber and Cellulosic Materials

Fiber and cellulosic materials, or lignocellulosic materials, are commonly found in agricultural, forestry, energy crops, and municipal paper and food waste. In Europe, especially Germany, a fair number of anaerobic digestion facilities utilize fiber and cellulosic materials as the feedstocks. Due to the intricate structural composition of lignocellulosic materials, their biodegradability is limited due to the slow hydrolysis rate. Consequently, the methane yield from such feedstock typically falls short of the theoretical value, often not exceeding 60% (Frigon & Guiot, 2010).

Lignocellulose, which forms the backbone of plant cell wall, consists primarily of cellulose (most abundant), hemicellulose, and lignin (Martínez et al., 2005; Sawatdeenarunat et al., 2015). Hemicellulos and cellulose are relatively easy to hydrolyze compared to lignin (Shrestha et al., 2017). Researchers have found that hydrolysis of cellulose and hemicellulose into fermentable sugars under anaerobic conditions is catalyzed by extracellular enzymes, which are released by hydrolytic bacteria (Maki et al., 2009). The general belief is that lignin cannot be hydrolyzed anaerobically as the cleavage of the aromatic rings in lignin requires oxygen (Dollhofer et al., 2015; Kirk et al., 1987). However, several recent studies have found evidence of lignin degradation in several natural anaerobic environments (Shrestha et al., 2017).

To overcome the slow hydrolysis rate of the lignocellulosic materials during anaerobic digestion, three strategies have been proposed to improve the hydrolysis efficiency: pretreatment, co-digestion, and inoculum selection and bioaugmentation. Various pretreatment methods, including physical (mechanical and thermal treatment), chemical (acid and alkaline treatment), physico-chemical, and biological (microbial and enzymatic treatment), have been investigated for lignocellulosic materials before anaerobic digestion to enhance their biodegradability (Shrestha et al., 2017). In contrast to physical and chemical pretreatment methods, biological pretreatment strategies are better suited for anaerobic digestion because they do not produce toxic compounds (e.g., phenolic compounds, furfurals, and hydroxymethylfurfural) that could impact the anaerobic digestion process (Frigon et al., 2012; Wu & He, 2013) and no steps are needed to neutralize strong acids or bases. Biological pretreatment using aerobic white-rot fungi has shown positive outcomes across various lignocellulosic feedstocks, including corn stover, wheat straw, soybean straw, switchgrass, and hardwood (Shrestha et al., 2017). In addition, hydrolytic enzyme pretreatment, combined with other pretreatment techniques (e.g., acid, alkaline, ultrasound, hydrothermal), has shown improved biogas production rate during anaerobic digestion (Bayr et al., 2013; Gomez-Tovar et al., 2012); however, the significant cost associated with hydrolytic enzymes poses a major challenge to the widely implement enzymatic pretreatment (Parawira, 2012).

Numerous studies have demonstrated that co-digestion of lignocellulosic materials with various co-substrates can significantly enhance methane production. Lignocellulosic substrates, which typically have high C/N ratio (Sawatdeenarunat et al., 2015), are commonly co-digested with nitrogen-rich substrates like animal manure to achieve a favorable C/N ratio of 20 to 30 for anaerobic digestion (Mata-Alvarez et al., 2014).

Choosing the right seed sludge type and concentration for digester's inoculation is critical as it is directly impacts initial performance, preventing slow startups, prolonged acclimation periods, and affecting long-term operational potential. To achieve a faster hydrolysis rate, previous research inoculated the anaerobic digester with rumen fluid, which provides a source of microbial populations with high hydrolytic activity, demonstrated a faster hydrolysis rate at the beginning (Shrestha et al., 2017). Moreover, adding specific microorganisms to a reactor to enhance a particular activity, i.e., bioaugmentation, has also been studied in recent years. Bioaugmentation with anaerobic bacteria and anaerobic fungi showed improved performance of digesting corn stover and corn silage, respectively (Nkemka et al., 2015; Zhang et al., 2015). However, it can be difficult to maintain microorganisms with one-time bioaugmentation due to wash-out and can be expensive to continuously supplement bioaugmentation cultures. To conquer the limitations, different types of reactors can be used to retain biomass and to separate the solids retention time from the hydraulic retention time, such as anaerobic membrane bioreactors and anaerobic sequencing batch reactors (Shrestha et al., 2017).

1.2.5 Other Macro and Micro Constituents

In anaerobic digestion, micronutrient supplementation can be required to support the microorganisms for optimum health. They are called micronutrients because very small quantities are required, typically in the nanogram per liter (ppb) concentration levels. Micronutrients are required for building new cellular material and act as cofactors in enzymes. Cofactors in enzymes support the catalytic action of enzymes. Without appropriate micronutrient supplement, methane production can still occur but may not be optimized. While micronutrients are important for methanogen archaea in the final methane production step, micronutrients can also benefit entire anaerobic systems by enhancing enzyme activity for the hydrolysis, acidogenesis, and acetogenesis steps of anaerobic digestion as well. Additionally, micronutrients can reduce competition to methanogenic archaea and help them to thrive in harsher environments.

1.2.5.1 Specific Micronutrients Required by Anaerobic Methanogen Archaea

The following are some of the specific micronutrients for anaerobic digestion anaerobic digestion and therefore co-digestion (Ariunbaatar et al., 2016):

- Iron
- Manganese
- Copper
- Zinc
- Selenium
- Nickel

- Cobalt
- Molybdenum

Iron is one of the most important micronutrients and promotes hydrolysis and acidification resulting in higher production of VFAs. Addition of iron can help to achieve higher volatile solids removal efficiency and help prevent methanogen resistance to VFA inhibition. This is because iron is key to enhancing the activity of certain metalloenzymes and helping methanogenic archaea quickly break down VFAs to prevent accumulation and inhibitory conditions. The importance of bioavailability and the specific type of iron present is discussed below in the paragraph on bioavailability. Beyond the specific metabolic benefits of iron, iron can also sequester sulfur in anaerobic reactor which can decrease the formation of hydrogen sulfide (H_2S), which can be inhibitory, corrosive, and malodorous (Bardi et al., 2023). Iron can begin to become inhibitory at does of 400 mg/L.

Zinc is crucial for hydrolysis (the breakdown of complex molecules into soluble organic molecules) and acidogenesis (the breakdown of soluble organic molecules to volatile fatty acids). However, zinc concentrations >50 mg/L can begin to be inhibitory on methane production. At concentrations >400 mg/L VFA production can be inhibited (Bardi et al., 2023).

Molybdenum is critical for synthesis of enzymes that reduce CO_2 and convert methyl compounds. Additionally, it can suppress activity of sulfate producing bacteria promoting a preference for the presence of methanogenic archaea. At concentrations above 1 mg/L, molybdenum can become toxic.

Small amounts of nickel, copper, cobalt, selenium, and tungsten are critical for the production of enzymes during methanogenesis, but high concentrations can quickly become inhibitory and toxic in the system. These metals both help in the synthesis and activation of metalloenzymes. Additionally, like iron, nickel addition has a benefit of promoting methane production in harsh environments such high levels of free ammonia. Cobalt is critical in the synthesis of vitamin B12 and corrinoids (Bardi et al., 2023).

1.2.5.2 Bioavailability

When supplementing anaerobic digestion and full scale, it is important to consider bioavailability. Many micronutrients, being metals, can bond to form insoluble compounds, making uptake by the organisms more difficult. Stressed or frequently changing anerobic environments can produce compounds that are chelating agents, which can bind to micronutrients making them less bioavailable (Bardi et al., 2023). This is very important for co-digestion systems where feedstocks can frequently change or have a wide variation. Fluctuating operating conditions such as pH, temperature hydraulic retention time (HRT), sludge retention time, organic loading rate (OLR), and mixing can stress the system, change the system biochemistry, and impact the bioavailability of microorganisms.

For iron, multiple forms are bioavailable and depending on form different microorganisms can benefit from the type of iron present in the system. Zero valent iron (ZVI) and Fe^{2+} facilitate help anaerobic archaea to survive in more harsh environments. Because iron is an electron donor ZVI and Fe^{3+} can help promote thermodynamics in hydrolysis, fermentation, and acetogenesis steps. Additionally, oxidized iron compounds such as magnetite and hematite can act as an electron exchange bridge (Bardi et al., 2023).

Certain copper salts can have a positive impact on digester performance resulting in increased biogas yields, such as copper sulfate ($CuSO_4$) and organic copper compounds. However, copper chloride ($CuCl_2$) has been found to have negative impacts on anaerobic digester performance, possibly due to its antibiotic effects. The organic copper compounds are found to be less bioavailable as they tend to precipitate in the preferred pH ranges for optimal anaerobic digestion (Bardi et al., 2023).

1.2.5.4 Co-digestion Micronutrient Benefits

One benefit of co-digestion is the addition of feedstocks introduces micronutrients not found in municipal sewer sludge. Co-digestion can decrease the amount of supplementation required, depending on the feedstock being introduced. Co-digestion with manure and municipal solid waste can provide the needed levels of iron to a deficient reactor primarily digesting cellulosic materials or silage. Livestock manure can contain high amounts of nickel, which can be beneficial to nickel deficient systems. Pig manure has a suitable amount of cobalt, making it a prime candidate for co-digestion, especially with food waste systems.

1.2.5.5 Co-digestion Micronutrient Concerns

In co-digestion, the incoming concentration of micronutrients can vary depending on feedstocks. Even with fairly consistent feedstocks, there can be inter- and intra-annual variation in micronutrient concentrations. Frequent monitoring is recommended to ensure optimum levels of micronutrients without allowing metals concentrations to increase to toxic levels. Additionally, if the final sludge application could be used for land application, the final concentrations of certain metals can exceed permitted application mass or concentration limits (Szaja et al., 2023).

Accounting for micronutrient concentrations when blending feedstock ratios is important for optimizing methane production. While co-digestion can be beneficial in providing the right mix of nutrients (macro and micro), certain co-substrates can rise to inhibitory and toxic levels. It is important to monitor influent concentrations of important micronutrients to optimize methane production while being careful to not exceed the inhibitory or toxic levels. Downstream effects of metals also need to be understood when considering the final disposition of the sludge (Szaja et al., 2023).

1.3 Important Feedstock Characteristics for Co-digestion

Feedstock characteristics are helpful for predicting biogas production (e.g., COD and biogas potential). The CHNO content can be a very powerful tool as it can be used to assess biodegradability of feedstock, predict changes in digester chemistry (pH, alkalinity, etc.), biogas production and even its composition.

1.3.1 Elemental Composition

Carbon is the key element driving methane potential in anaerobic digestion. Higher carbon density results in greater methane potential from a feedstock. However, this is not the only factor, as carbon contributes to both methane (CH_4) and carbon dioxide (CO_2) respiration. Depending on the carbon source and other factors, $CH_4:CO_2$ will vary.

Hydrogen is the other key element for methane potential. While most methane will be produced from acetate (CH_3COO), hydrogen gas (H_2) is formed as a byproduct of several metabolic pathways and is directly combined with CO_2 to generate methane. As such, higher hydrogen ratio will generally improve $CH_4:CO_2$.

Nitrogen does not directly contribute to methane production but is important to maintaining digester health. It supports microbial growth and is the main source of alkalinity in the form of ammonia (NH_4), but at higher concentrations ammonia concentrations can be inhibitory to methane production. The ratio of carbon to nitrogen, C:N, should be targeted at 20–30 for optimal biogas production (Li et al., 2011).

Oxygen will be present in many organic compounds, particularly carbohydrates and proteins. In anaerobic digestion oxygen will be metabolized to produce either H_2O or CO_2, reducing the carbon and hydrogen available for methane (CH_4). The greater the oxygen ratio in a substrate, the lower the expected $CH_4:CO_2$.

Conventional anaerobic digestion feedstocks from municipal waste are a mixture of complex biological compounds. This contributes to a complex elemental composition. For co-digestion feed stocks however, the elemental composition is more uniform. Feed stocks for co-digestion come from a variety of sources and can be broken into three main categories—carbohydrates, lipids, and proteins.

Carbohydrates—$C_nH_{2n}O_n$, for example, $C_6H_{12}O_6$ (Glucose)
Lipids—$CH_3(CH_2)_nCOOH$, for example, $CH_3(CH_2)_{16}COOH$ (Stearlic acid)
Proteins—$C_{10}H_{20}O_6N_2$ (empirical)

In addition to the elemental compositions listed, there will be some additional contribution of sulfur and phosphorus.

Once the elemental composition of the substrates is known, co-digestion performance can be optimized by using different combinations and types of substrates, so that the digester can be provided with the right amounts of carbon, nitrogen, and micro- and macronutrients. Though these values vary in the literature, they can be used as general guidelines to determine a suitable feedstock.

1.3.2 Total and Soluble Chemical Oxygen Demand

Chemical oxygen demand is the measurement of oxygen consumption during decomposition of organic matter via a reaction with oxidizing agent. In anaerobic digestion, COD is an important indicator of organic content that can be degraded through the anaerobic process, and it is a key parameter used for operation and performance evaluation.

Total COD (tCOD) represents the oxygen demand for both soluble and non-soluble (particulate) fraction, while soluble COD (sCOD) measures the oxygen demand for only the soluble fraction of the sample. Testing for COD involves determining the oxidant amount that reacts with a sample under controlled environment, using a strong chemical oxidizing agent, such as potassium dichromate or potassium permanganate. When measuring sCOD, the sample is initially filtered through a 0.45-mm filter before analysis to determine the oxidant demand.

1.3.3 Total Solid and Volatile Solid

Total solid (TS) is measurement of total mass of solids present in a sample, while volatile solid (VS) represents only the organic portion of the sludge that is combustible. Measurement of TS and VS can be performed using EPA Method 1684 (U.S. EPA, 2001). TS is measured by drying the sample at 105 °C and is calculated as below:

$$\%\ total\ solids = \frac{W_{total} - W_{dish}}{W_{sample} - W_{dish}} \times 100$$

Or

$$\frac{mg\ \%\ total\ solids}{kj\ sludge} = \frac{W_{total} - W_{volatile}}{W_{sample} - W_{dish}} \times 1\ 000\ 000$$

Where:

W_{dish} = Weight of dish (mg),
W_{total} = Weight of wet residue and dish (mg), and
W_{sample} = Weight of dried residue and dish after ignition (mg).

Volatile solids is measured by combusting the sludge sample at 550 °C for two hours and is calculated as below:

$$\%\ volatile\ solids = \frac{W_{total} - W_{volatile}}{W_{total} - W_{dish}} \times 100$$

Or

$$\frac{mg\ volatile\ solids}{kj\ sludge} = \frac{W_{total} - W_{volatile}}{W_{total} - W_{dish}} \times 1\ 000\ 000$$

Where:

W_{total} = Weight of dish (mg),
W_{total} = Weight of dried residue and dish (mg), and
W_{total} = Weight of residue and dish after ignition (mg).

1.3.4 Biomethane Potential

Most high-strength waste typically has higher methane potentials than municipal wastewater sludge. Fats/lipids have a theoretical yield of 16 cu ft CH_4/lb $VS_{destroyed}$ compared to 6.6 cu ft/lb $VS_{destroyed}$ for protein and 10 cu ft/lb $VS_{destroyed}$ for carbohydrates (Long et al., 2012).

Typically, increase in the methane potential is due to a synergistic effect between the sludges and the feedstocks that results in a greater methane generation while co-digested than if it were digested separately (Higgins et al., 2017).

The synergistic effect is defined as a phenomenon where the co-digestion of substrates with sewage sludge results in greater methane generation than when the substrates are digested separately. These effects have been reported in several studies for lab- and bench-scale systems as well as full-scale systems (Aichinger et al., 2015; Labatut et al., 2011; Zhang et al., 2015). Full-scale, pilot-scale, and lab-scale studies of co-digestion of FOG, dairy waste, manure, and other HSWs with municipal sewage sludge have been conducted across the world (Kabouris et al., 2009; Suto et al., 2006).

Biomethane potential is expressed in the volume of biogas that is estimated from destruction of a unit mass of volatile solids. The typical range of values for BMP is 12 to 20 cf/lb-VS-destroyed. Table 3.2 provides a review of BMP data for various feedstocks.

Factors that can affect BMP include the amount of volatile solids that are available for destruction within the material. For example, primary waste has greater potential for volatile solids reduction than secondary

TABLE 3.2 Biomechanical Methane Potential for Various Feedstocks

Feedstock	Digestion Type	Loading (lb VS/ cu ft/day)	BMP (cu ft/lb VS_{added})	VSR of the Feedstock (%)	Reference
Skin trimmings, flesh, and tannery wastewater sludge	T	0.25	9.61	N/D	(Zupančič & Jemec, 2010)
Solid slaughter house waste	M	0.056	8.3–8.8	N/D	Salminen & Rintala, 2002)
Domestic swill refuse (full-scale)	M	0.06	9.6	81% (SS + swill)	(Zupančič et al., 2008
Cheese whey	M	0.1	4.16	N/D	(Bertin et al., 2013)
Sewage sludge and brewery waste	M	0.05–0.09	8.2–9.6*	16–19 (different ratios of sludge and brewery waste by wet weight)	(Pecharaply et al., 2007)
Dairy, manure, FW, fat from fish canning	M	N/A	3.6–6.6	49.1 ± 5.7–52.1 ± 5.0 (ratios of cow manure, FW, and fish fat)	(Neves, Oliveira, & Alves, 2009; Neves et al., 2009)
Brewery Waste	M	0.14	10*		Rajagopalan et al., 2022
Creamery and Poultry Waste	M	0.08–0.16	17*		
Blended Whey	M	0.095	16.04*		

Note: *On basis of $VS_{removed}$ instead of VS_{added} as in the other cases.

waste, as the secondary waste has already been subjected to biological degradation as part of the WRRF process.

1.3.5 Heavy Metal Content

Trace levels of heavy metals can be essential micronutrients for the anaerobic digestion process and have been shown to be important elements to carry out several cellular functions, such as supporting diverse enzymatic reactions and prohibiting sulfide toxicity (Kadam et al., 2022). Some specific examples are cobalt, iron, and nickel that play critical roles in electron transport catabolism, whereas large amounts of iron, nickel, and cobalt and some molybdenum and zinc have been shown essential to develop enzymes for the methanogenesis pathway (Glass & Orphan, 2012).

High levels of heavy metal content in substrates can also potentially cause perturbation and reduction of the co-digestion process by impacting the methanogenic and acetogenic community, with higher impacts on the former (Alrawashdeh, 2022). Heavy metals with toxicity to anerobic digestion process include copper, zinc, chromium, and lead, ordered in the level of relative toxicity with respect to methane content, with copper being the most toxic and lead the least toxic (Alrawashdeh, 2022; Nguyen et al., 2019). With respect to removal of TS and biogas yield, similar relative toxicity was observed for lead (least toxicity) and copper (highest toxicity), but with zinc and chromium showing similar levels of toxicity (medium toxicity) (Alrawashdeh, 2022).

Following co-digestion, the heavy metal content splits into biosolids and digestate, and remaining residuals require management. The approach for management of biosolids is governed federally by 40 CFR Part 503 (Part 503) which defines Class A (Exceptional Quality) and Class B biosolids. It should be noted that some states have applied stricter limits for pathogen reduction and heavy metals concentrations relative to Part 503.

1.3.6 Nitrogen and Phosphorus Content

High nitrogen levels and maybe phosphorus to a lesser extent are components of concern mainly because of the increased nutrient content in dewatering recycle streams post co-digestion. Recycled streams tend to be highly concentrated and increase the nutrient load on secondary treatment and also raise the final effluent nutrient concentrations impacting regulatory limits of utilities. The other issue of high nitrogen content feedstock is the potential for ammonia inhibition in the digester. Ammonia inhibition is directly related to the carbon–nitrogen (C/N) ratio of the feedstock. It has been popularly accepted that optimal co-digestion is about balancing the C/N ratio in the co-substrate mixture, as well as the macro- and micronutrients and the pH in the digester (Higgins et al., 2017). The C/N ratio of the substrates is an important indicator of performance. High C/N ratios can deplete the necessary N source for biomass growth, whereas lower C/N ratios lead to the accumulation of ammonia. For stable operation, the optimum C/N ratio falls between 20 and 70 g COD/g N (Mata-Alvarez et al., 2014).

A wide range of nitrogen concentrations has been reported for different feedstocks ranging from about 0.6% to 5% dw (Rauch-Williams et al., 2023). Typically, N loads are larger compared to additional phosphorus loads from feedstock (Rauch-Williams, 2023) and therefore some utilities may need side stream N treatments to meet final effluent limits.

1.3.7 Soluble Cation Concentrations

Soluble cation concentrations, namely Ca_2^+, Mg_2^+, Na^+, K^+, and NH_4^+ are impacted due to co-digestion, due to the addition of feedstocks of different elemental concentrations. Soluble cations influence the monovalent to divalent cation (M/D) ratio, which is an important indicator of dewatering efficiency. In general, as the M/D increases above two, polymer demand and cake solids worsen (Higgins & Murthy, 2006; Murthy & Novak, 1999; Novak, 2006). Several studies have demonstrated the addition of certain co-wastes improved dewatering considerably. Co-digestion of FOG with municipal sludge decreased the volume of biosolids (reduction in the wet cake produced) by 33% and improved biosolids dewaterability. Total polymer also reduced by 11% (York & Magner, 2009). Full-scale whey co-digestion indicated no net increase in the amount of cake solids produced and the polymer demand decreased by 25% (Aichinger et al., 2015).

Specifically, magnesium concentrations in feedstock are of importance given its impact on struvite formation. Co-digestion of feedstock with very high Mg levels may increase struvite formation in the digesters and downstream components.

1.4 Sampling Protocols for Feedstock Characteristics

Most feedstocks need to be evaluated for TS, VS, COD, and nutrients on a regular basis. Typically, most digester feed and effluent testing is performed according to *Standard Methods for the Examination of Water and Wastewater* (American Public Health Association et al., 2012) or specific protocols developed for laboratory testing, as discussed in this section.

1.4.1 Pre-processed and Polished Feedstock

If there are concerns about incoming new feedstock, an anaerobic toxicity test should be performed. Alternatively, a BMP/respirometry test could be conducted to estimate gas production from the feedstock material. The most information about the feedstock and its behavior in a digester can be obtained from running a pilot digester. It is most beneficial to conduct the BMP/respirometry/pilot digestion with the sludge/feedstock combination that would be co-digested in full-scale, in a VS loading rate as close as possible to the real-life facility scenario.

Other important parameters that should be tested for in a feedstock include toxic compounds such as acids, ammonia, total cyanide, and heavy metals. In the case of waste streams from industrial sources, other specific chemicals used by the specific industrial process (for instance use of strong alkalis/acids in food process wastewater) should also be determined as they can possibly accumulate in the digester with time and might inhibit the anaerobic digestion process.

The composition of a given feedstock may not be constant, it can change according to variations in its supply and production. As a result, samples should be tested over a period of time. Constant communication with the feedstock supplier/processor is key so that deficiencies in feedstock quantity and quality can be identified by the utility.

It is important that the analyses be carried out for both the sludges and feedstock as well as digester contents and digester biogas as listed in Table 3.3. This is important as it presents the overall condition of digestion when feedstocks are used.

1.4.2 Feedstock Quality Criteria for Co-digestion

In the United States, while there are no direct regulations governing the feedstock quality for co-digestion, utilities will require complying with their water quality, air quality, and solids management permits that can be impacted by implementing co-digestion. There may be a few potential feedstock quality criteria that may indirectly relate to the WRRF permitting, as discussed below.

Some utilities that have been practicing co-digestion for a while have developed feedstock specifications they negotiate with their suppliers, taking into account their specific regulations, and operational requirements (Rauch-Williams et al., 2023). Bioslurry feedstock manufacturers have developed quality standards they provide to their utility customers.

The following are some of the things to consider when choosing co-digestion feedstocks:

1. Local availability: Though not a quality criterion, proximity to the WRRF can ensure that the feedstock is not stored for longer periods of time resulting in odor/emissions/fermentation of certain feedstocks.
2. Moisture and total solids: Typically, raw feedstock needs to be pre-processed and diluted before being fed to the digester. Operators need

TABLE 3.3 Analyses for Feedstock Evaluation

Parameter	Raw Feedstock	Pretreated/Polished Feedstock	Digester Contents	Testing Method	Reason for Test
Visual inspection	X	X			Check for consistency and contamination.
CHNOS	X	X	X	Elemental analyzer	Elemental composition
COD	X	X	X	HACH test kit	Strength of the material can be used to determine organic loading rates (OLRs, units of lb/ cu ft·d or kg/m^3·d) and COD removal efficiencies (%).
BOD	X	X	X	Standard Methods	Strength of the material
TS/VS	X	X	X	Standard Methods	TS/VS ratio is a measurement of strength. TS measurement (mg/L or %) is most frequently used for reporting the solids concentration of sludge. Can be used to determine organic loading rates (OLRs, units of lb/cu ft·d or kg/m^3·d) and VS removal efficiencies (%).
TSS/VSS	X	X	X	Standard Methods	
pH	X	X	X	Standard Methods	Makes sure feedstock is not overly acidic or basic. Indicator of stability in digester.
Volatile acids			X	HACH test kit/SM 5560 C	Indicator of stability in digester.
Total alkalinity			X	SM 2320 B	Indicator of stability in digester.

NH_3-N	X	X	X	Standard Methods/ Ammonia Probe	Essential nutrient/ in excess an inhibitor. Need estimate for for centrate return streams to secondary.
Conductivity	X			Conductivity Probe	High conductivity can affect the potential for reuse
Sulfides/chlorides	X			Standard Methods	For regulatory purposes; High salt content can affect the potential for reuse
TP/OP	X	X	X	Standard Methods/ HACH kit	Essential nutrient/ in excess an inhibitor. Need estimate for centrate return streams to secondary.
K	X	X	X	HACH test kit/	Potential inhibition of methanogens
Fe, Mg, Ca, and Na	X	X	X	GC-MS	
Heavy metals and toxicity scan	X	X		EPA 9014, Total Cyanide TTLC CAM 17 Metals, STLC (for industrial feedstock)	Potential inhibition for digestion/toxicity
Biogas quantity			X	Flow Meters	Biogas formation potential is tracked to correlate to the feedstock
Biogas composition			X	Draeger tubes	Indicator of acclimation. Under steady-state operating conditions, average CH_4 content is expected to be 50%–70%. A decrease in CH_4 concentration below 50% may indicate a potential system upset.

to ensure that the TS content allows ease of pumping and mixing with the current equipment.

3. Mixing and particle size: It is important to have a well-mixed digester. Based on the source and type of co-digestion feedstock, the material or particle size can vary widely. Smaller particle size has been shown to increase biogas yield due to better contact between the methanogens and the feed. In most pretreatments for feedstock, shredding or grinding is carried out. It is optimal to accept such feedstock. Further description of feedstock pretreatment is discussed in Chapter 5.
4. Toxicity: The feedstock must not be toxic to the methanogens or produce intermediates or release gases that are toxic. If feed has a high protein content, during digestion, the breakdown of the amino acids present in the protein will produce ammonia gas and hydrogen sulfide, which can inhibit digestion.
5. Nutrient balance (C:N): The optimal C/N ratio in methane fermentation is between 25 and 30 (Nurliyana et al., 2015).
6. Permitting: While there are no direct regulations governing feedstock quality, a holistic picture of the permits for the liquid, gas, and solids streams need to be evaluated for both quality and quantity. The nutrients in the centrate stream post-digestion are recycled to the secondary treatment and may impact the utility meeting final discharge limits for N. Local air permitting authorities may also look into the increase in methane production and H_2S emissions. Odor emissions in the utility's feedstock receiving facilities may need to be evaluated as a cause for air permitting concerns. With regard to the solids streams, while co-digestion may increase biosolids production, it typically does not affect the ability to meet regulatory requirements related to biosolids quality.
7. Biodegradability, gas production, and energy production: One of the most important parameters to consider is if the feedstock is easily digestible under the conditions of the utility's digester and will produce more energy biogas. Conversion of this biogas into energy and energy utilization are bigger decisions and beyond the scope of this chapter, but having an energy positive feedstock is necessary.
8. pH: The necessary pH for anaerobic digestion ranges somewhere between 6.8 to 8.5. Most FOG and other feedstock are lower in pH. Well-buffered digesters will be able to function well even with lower pH feedstocks if loaded at a reasonable rate. If the feedstocks are too basic or acidic, they may not be suited for co-digestion.

9. Institutional support and societal acceptance: Understanding the competition among various industries as well as how the public would receive the byproduct are important considerations when evaluating a feedstock. This may also affect the perceived value of the byproduct.

1.4.3 *Best Practices for Feedstock Evaluation*

Water resource recovery facilities could set up minimum characteristics for feedstock acceptance:

1. Contract with waste supplier to include contaminants to be avoided.
2. Screen deliveries for TS, VS, and visual consistency.
3. Require minimum strength of feedstock.
4. Limit ammonia in feedstock.
5. Require representative sample at loading time to analyze in-house.

1.5 Feedstock Composition and Relationship to Digester Performance

Determining the composition of feedstock can be helpful to predict the fate and behavior of the elemental compounds during anaerobic digestion. For example, the VS reduction and the subsequent biogas yield of a feedstock can be predicted if the composition of fats, proteins, and carbohydrates, and the CHNO content is known. More detailed information in this regard is presented in Chapter 4 of this manual.

1.5.1 *Stoichiometric Models*

The disintegration and hydrolysis steps of anaerobic digestion are best represented using first-order kinetics (reference needed), with the product being monosaccharides, amino acids, and long-chain fatty acids. From these monomers, stoichiometric models of the biological processes which comprise acidogenesis, acetogenesis, and methanogenesis can be used to estimate methane yield and other digestion characteristics.

Acidogenesis/Acetogenesis

Monosaccharides

$$C_6H_{12}O_6 + 2H_2O \rightarrow 2CH_3COOH + 2CO_2 + 4H_2$$

Amino acids

$$C_{10}H_{20}O_6N_2 \rightarrow 5.5CH_3COOH + 4.5CO_2 + 2NH_3 \text{ (empirical)}$$

$C_3H_7NO_2 + 2C_2H_5NO_2 + 2H_2O \rightarrow 3CH_3COOH + CO_2 + 3NH_3$ (Strickland fermentation)

Long-chain fatty acids

$$CH_3(CH_2)_{16}COOH + 16H_2O \rightarrow 9CH_3COOH + 16H_2 \text{ (}\beta\text{-oxidation)}$$

Methanogenesis

Aceticlastic

$$CH_3COOH \rightarrow CH_4 + CO_2$$

Hydrogen utilizing

$$4H_2 + CO_2 \rightarrow CH_4 + H_2O$$

Chapter 4 provides more detail on stoichiometric modeling.

1.5.2 Process Modeling

Several models for anaerobic digestion have been developed to assist in the design and operation of anaerobic digesters. This is of interest when planning for the implementation of a co-digestion program. The ability to model the impacts of different feed stock combinations and other operational changes can improve digester performance and reduce the risk of process upset. However, many models are sensitive to the accuracy of input data and may require significant tuning using physical data to produce accurate predictive results.

1.5.2.1 Anaerobic Digestion Model No. 1

Anaerobic Digestion Model No. 1 was published by IWA in 2002 and represents a comprehensive effort to create a generic model of the anaerobic digestion process. The ADM1 model comprises a set of dynamic state variables representing soluble and particulate substrate concentrations, with the accompanying kinetic parameters and rates. It can be implemented in either differential-algebraic equation or differential equation forms (Batstone et al., 2002).

1.5.2.2 Anaerobic Digestion Model No. 2

Anaerobic Digestion Model No. 2 was developed by the Institute of Agronomic Research (INRA) in France. It accommodates a wider range of substrates, more microbial groups and interactions, and uses more detailed kinetic equations. Additionally, ADM2 is more complex than its predecessor ADM1, which means it requires a larger number of parameters to be estimated. Parameter estimation needs extensive experimental data gathering.

2.0 REFERENCES

Adnane, I., Taoumi, H., Elouahabi, A., Lahrech, K., & Oulmekki, A. (2024). Valorization of crop residues and animal wastes: Anaerobic co-digestion technology. *Heliyon*, e26440–e26440. https://doi.org/10.1016/j.heliyon.2024.e26440

Afazeli, H., Jafari, A., Rafiee, S., & Nosrati, M. (2014). An investigation of biogas production potential from livestock and slaughterhouse wastes. *Renewable and Sustainable Energy Reviews*, *34*, 380–386.

Ahmad, B., Yadav, V., Yadav, A., Rahman, M. U., Yuan, W. Z., Li, Z., & Wang, X. (2020). Integrated biorefinery approach to valorize winery waste: A review from waste to energy perspectives. *Science of the Total Environment, 719*, 137315.

Aichinger, P., Wadhawan, T., Kuprian, M., Higgins, M., Ebner, C., Fimml, C., Murthy, S., & Wett, B. (2015). Synergistic co-digestion of solid-organic-waste and municipal-sewage-sludge: 1 plus 1 equals more than 2 in terms of biogas production and solids reduction. *Water Research, 87*, 416–423.

Alrawashdeh, K. A. (2022). Anaerobic co-digestion efficiency under the stress exerted by different heavy metals concentration: An energy nexus analysis. *Energy Nexus, 7*, 100099.

American Public Health Association, American Water Works Association, & Water Environment Federation. (2012). *Standard methods for the examination of water and wastewater* (22nd ed.). E. W. Rice, R. B. Baird, A. D. Eaton, & L. S. Clesceri, Eds. American Public Health Association.

Ariunbaatar, J., Esposito, G., Yeh, D. H., & Lens, P. N. L. (2016). Enhanced anaerobic digestion of food waste by supplementing trace elements: Role of selenium (VI) and iron (II). *Frontiers in Environmental Science*, *4*(8). https://doi.org/10.3389/fenvs.2016.00008

Atelge, M. R., Atabani, A. E., Rajesh Banu, J., Krisa, D., Kaya, M., Eskicioglu, C., Kumar, G., Lee, C., Yildiz, Y. Ş., Unalan, S., Mohanasundaram, R., & Duman, F. (2020). A critical review of pretreatment technologies to enhance anaerobic digestion and energy recovery. *Fuel*, *270*, 117494–117494. https://doi.org/10.1016/j.fuel.2020.117494

Bachmann, S. A. L., Calvete, T., & Féris, L. A. (2022). Potential applications of brewery spent grain: Critical an overview. *Journal of Environmental Chemical Engineering, 10*(1), 106951.

Bardi M., Vinardell S., Astals S., & Koch, K. (2023). Opportunities and challenges of micronutrients supplementation and its bioavailability in anaerobic digestion: A critical review. *Renewable and Sustainable Energy Reviews, 186*.

Bareha, Y., Girault, R., Jimenez, J., & Tr´emier, A. (2018). Characterization and prediction of organic nitrogen biodegradability during anaerobic digestion: A bioaccessibility approach. *Bioresource Technology, 263* (February), 425–436.

Batstone, D., Keller, J., Angelidaki, I., Kalyuzhnyi, S., Pavlostathis, S., Rozzi, A., Sanders, W., Siegrist, H., & Vavilin, V. (2002). Anaerobic digestion model No 1 (ADM1). *Water Science and Technology: A Journal of the International Association on Water Pollution Research, 45*, 65–73.

Bayr, S., Kaparaju, P., & Rintala, J. (2013). Screening pretreatment methods to enhance thermophilic anaerobic digestion of pulp and paper mill wastewater treatment secondary sludge. *Chemical Engineering Journal, 223*, 479–486. https://dx.doi.org/10.1016/j.cej.2013.02.119

Bertin, L., Grilli, S., Spagni, A., & Fava, F. (2013). Innovative two-stage anaerobic process for effective codigestion of cheese whey and cattle manure. *Bioresource Technology, 128*, 779–783.

Bochmann, G., Gronauer, A., & Weiss, S. (2015). *Biogas science and technology*. Springer.

Cao, F., Wu, Y., Xu, L., Song, X., & Ding, J. (2024). Microbial community changes and metabolic pathways analysis during waste activated sludge and meat processing waste anaerobic co-digestion. *Journal of Environmental Management*. https://doi.org/10.1016/j.jenvman.2024.121444

Chiappero, M., Berruti, F., & Fiore, S. (2023). Biomethane potential of wine lees from mesophilic anaerobic digestion. *Biochemical Engineering Journal, 196*, 108954.

Cirne, D. G., Paloumet, X., Björnsson, L., Alves, M. M., & Mattiasson, B. (2007). Anaerobic digestion of lipid-rich waste—Effects of lipid concentration. *Renewable Energy, 32*(6), 965–975.

Coyne, J., Wilson, C., Scarborough, M., & Umble, A. (2017). *Anaerobic digestion fundamentals fact sheet*. Water Environment Federation, Municipal Resource Recovery Design Committee.

da Ros, C., Cavinato, C., Cecchi, F., & Bolzonella, D. (2014). Anaerobic co-digestion of winery waste and waste activated sludge: Assessment of process feasibility. *Water Science & Technology, 69*(2), 269–277.

da Ros, C., Cavinato, C., Pavan, P., & Bolzonella, D. (2014). Winery waste recycling through anaerobic co-digestion with waste activated sludge. *Waste Management, 34*(11), 2028–2035.

de los Reyes III, F. L., Portiolli, G., & Guertin, E. (2022). *Developing high-yield and resilient anaerobic co-digestion of alternative wastes* [Grant report]. Environment Research and Education Foundation.

Deng, Z., Ferreira, A. L. M., Spanjers, H., & van Lier, J. B. (2023). Anaerobic protein degradation: Effects of protein structural complexity, protein concentrations, carbohydrates, and volatile fatty acids. *Bioresource Technology Reports, 22*, 101501.

Dollhofer, V., Podmirseg, S. M., Callaghan, T. M., Griffith, G. W., & Fliegerová, K. (2015). Anaerobic fungi and their potential for biogas production. In: G. M. Gübitz, A. Bauer,

Eleutheria, N., Iossifidou, M., Tataki, V., Eftaxias, A., Aivasidis, A., & Diamantis, V. (2016). Energy recovery and treatment of winery wastes by a compact anaerobic digester. *Waste and Biomass Valorization,* 799–805.

Fabbri, A., Bonifazi, G., & Serranti, S. (2015). Micro-scale energy valorization of grape marcs in winery production plants. *Waste Management, 36*, 156–165.

Frigon, J.-C., & Guiot, S. R. (2010). Biomethane production from starch and lignocellulosic crops: A comparative review. *Biofuels, Bioproducts and Biorefining, 4*(4), 447–448. https//doi.org/10.1002/bbb.229

Frigon, J. C., Mehta, P., & Guiot, S. R. (2012). Impact of mechanical, chemical and enzymatic pre-treatments on the methane yield from the anaerobic digestion of switchgrass. *Biomass Bioenergy, 36*, 1–11. http://dx.doi.org/10.1016/j.biombioe.2011.02.013

Glass, J. B., & Orphan, V. J. (2012). Trace metal requirements for microbial enzymes involved in the production and consumption of methane and nitrous oxide. *Frontiers in Microbiology, 61*.

Goberna, M., del Mar Camacho, M., Lopez-Abadia, J. A., & García, C. (2013). Co-digestion, biostimulation and bioaugmentation to enhance methanation of brewer's spent grain. *Waste Management & Research, 31*(8), 805–810.

Gomez-Tovar, F., Celis, L.B., Razo-Flores, E., & Alatriste-Mondragón, F. (2012). Chemical and enzymatic sequential pretreatment of oat straw for methane production. *Bioresource Technology, 116*, 372–378. https://doi.org/10.1016/j.biortech.2012.03.109

Higgins, M., & Murthy, S. (2006). *Understanding factors affecting polymer demand for thickening and dewatering* (WERF report). IWA Publishers.

Higgins, M., Rajagopalan, G., Miller, A., Brown, J., & Beightol, S. (2017). Effect of food waste co-digestion on digestion, dewatering, and cake quality. *Water Environment Research*, *89*(1), 32–42.

Higgins, M., Van Horne, M., Loomis, P., & Parry, D. (2017, November 2). Anaerobic digestion 101 [Webinar]. Water Environment Federation, Plant Operations and Maintenance Community. https://www.wef.org/globalassets/assets-wef/2-resources/online-education/webcasts/presentation-handouts/presentation-handouts---advanced-digestion-from-operators-perspective-11-2-17.pdf?_t_ip=40.77.167.33

Holohan, B. C., Duarte, M. S., Szabo-Corbacho, M. A., Cavaleiro, A. J., Salvador, A. F., Pereira, M. A., Ziels, R. M., Frijters, C. T. M. J., Pacheco-Ruiz, S., Carballa, M., Souuza, D. Z., Stams, A. J. M., O'Flaherty, V., van Lier, J. B., & Alves, M. M. (2022). Principles, advances, and perspectives of anaerobic digestion of lipids. *Environmental Science & Technology, 56*, 4749–4775.

Jiang, Y., McAdam, E., Zhang, Y., Heaven, S., Banks, C., & Longhurst, P. (2019). Ammonia inhibition and toxicity in anaerobic digestion: A critical review. *Journal of Water Process Engineering, 32*, 100899.

Kabouris, J. C., Tezel, U., Pavlostathis, S. G., Engelmann, M., Dulaney, J., Gillette, R. A., & Todd, A. C. (2009). Methane recovery from the anaerobic codigestion of municipal sludge and FOG. *Bioresource Technology, 100*(15), 3701–3705.

Kadam, R., Khanthong, K., Jang, H., Lee, J., & Park, J. 2022. Occurrence, fate, and implications of heavy metals during anaerobic digestion: A review. *Energies, 15(*22), 8618.

Kassongo, J., Shahsavari, E., & Ball, A. S. (2020). Co-digestion of grape marc and cheese whey at high total solids holds potential for sustained bioenergy generation. *Molecules*, *25*(23), 5754. https://doi.org/10.3390/molecules25235754

Kirk, T. K., Gifford, O., Drive, P., & Farrell, R. L. (1987). Enzymatic "combustion": The microbial degradation of lignin. *Microbiology*, 465–505.

Labatut, R. A., Angenent, L. T., & Scott, N. R. (2011) Biochemical methane potential and biodegradability of complex organic substrates. *Bioresource Technology, 102*(3), 2255–2264.

Li, Y., Park, S. Y., & Zhu, J. (2011). Solid-state anaerobic digestion for methane production from organic waste. *Renewable and Sustainable Energy Reviews, 15*(1), 821–826.

Liu, Y., Ngo, H. H., Guo, W., Peng, L., Wang, D., & Ni, B. (2019). The roles of free ammonia (FA) in biological wastewater treatment processes: A review. *Environment International, 123,* 10–19.

Long, J. H., Aziz, T. N., & Ducoste, J. J. (2012). Anaerobic co-digestion of fat, oil, and grease (FOG): A review of gas production and process limitations. *Process Safety and Environmental Protection, 90*(3), 231–245.

Maki, M., Leung, K. T., & Qin, W. (2009). The prospects of cellulase-producing bacteria for the bioconversion of lignocellulosic biomass. *International Journal of Biological Sciences, 5*, 500–516. http://dx.doi.org/10.7150/ijbs.5.500

Mallen, E., & Najdanovic-Visak, V. (2018). Brewers' spent grains: Drying kinetics and biodiesel production. *Bioresource Technology Reports, 1*, 16–23.

Martínez, Á. T., Speranza, M., Ruiz-Dueñas, F. J., Ferreira, P., Camarero, S., Guillén, F., Martínez, M. J., Gutiérrez, A., & Del Río, J. C. (2005). Biodegradation of lignocellulosics: Microbial, chemical, and enzymatic aspects of the fungal attack of lignin. *International Microbiology, 8*(3), 195–204.

Mata-Alvarez, J., Dosta, J., Romero-Güiza, M. S., Fonoll, X., Peces, M., & Astals, S. (2014). A critical review on anaerobic co-digestion achievements between 2010 and 2013. *Renewable and Sustainable Energy Reviews, 36*, 412–427. http://doi.org/10.1016/j.rser.2014.04.039

Mathias, T. R. S., Alexandre, V. M. F., Cammarota, M. C., de Mello, P. P. M., & Sérvulo, E. F. C. (2015). Characterization and determination of brewer's solid wastes composition. *Journal of the Institute of Brewing, 121*(3), 400–404.

McInerney, M. J. (1988). Anaerobic hydrolysis and fermentation of fats and proteins. In *Biology of Anaerobic Organisms* (pp. 373–416). John Wiley & Sons.

Morales-Polo, C., Cledera-Castro, M. D. M., & Moratilla Soria, B. Y. (2018). Reviewing the anaerobic digestion of food waste: From waste generation and anaerobic process to its perspectives. *Applied Sciences, 8*(10), 1804.

Murthy, S. N., & Novak, J. T. (1999). Factors affecting floc properties during aerobic digestion: Implications for dewatering. *Water Environment Research, 71*(2), 197–202.

Muszyński-Huhajło, M., Zięba, B., Janiak, K., Miodoński, S., Jurga, A., & Szetela, R. (2021). Can nitrifiers from the sidestream deammonification process be a remedy for the N-overload of the mainstream reactor? *Science of the Total Environment, 790*(10), 148066. https://doi.org/10.1016/j.scitotenv.2021.148066

Nazifa, T. H., Saady, N. M. C., Bazan, C., Zendehboudi, S., Aftab, A., & Albayati, T. M. (2021). Anaerobic digestion of blood from slaughtered livestock: A review. *Energies, 14*(18), 5666. https://doi.org/10.3390/en14185666

Neves, L., Oliveira, R., & Alves, M. M. (2009). Co-digestion of cow manure, food waste and intermittent input of fat. *Bioresource Technology*, *100*(6), 1957–1962.

Neves, L., Pereira, M. A., Mota, M., & Alves, M. M. (2009). Detection and quantification of long chain fatty acids in liquid and solid samples and its relevance to understand anaerobic digestion of lipids. *Bioresource Technology*, *100*(1), 91–96.

Nguyen, Q-M., Bui, D-C., Phuong, T., Doan, V-H., Nguyen, T-N., Nguyen, M-V., Tran, T-H., & Do, Q-T. (2019). Investigation of heavy metal effects on the anaerobic co-digestion process of waste activated sludge and septic tank sludge. *International Journal of Chemical Engineering, 1*, 5138060. https://doi.org/10.1155/2019/5138060

Nkemka, V. N., Gilroyed, B., Yanke, J., Gruninger, R., Vedres, D., McAllister, T., & Hao, X. (2015). Bioaugmentation with an anaerobic fungus in a two-stage process for biohydrogen and biogas production using corn silage and cattail. *Bioresource Technology, 185,* 79–88. https://doi.org/10.1016/j.biortech.2015.02.100

Novak, J. T. (2006). Dewatering of sewage sludge. *Drying Technology, 24*(10), 1257–1262.

Noyola A., & Tinajero A. (2005). Effect of biological additives and micronutrients on the anaerobic digestion of physicochemical sludge. *Water Science and Technology, 52*(1–2), 275–281.

Nurliyana, M. Y., H'ng, P. S., Rasmina, H., Kalsom, M. U., Chin, K. L., Lee, S. H., Lum, W. C., & Khoo, G. D. (2015). Effect of C/N ratio in methane productivity and biodegradability during facultative co-digestion of palm oil mill effluent and empty fruit bunch. *Industrial Crops and Products, 76,* 409–415.

Parawira,W. (2012). Enzyme research and applications in biotechnological intensification of biogas production. *Critical Reviews in Biotechnology, 32*, 172–186. https://doi.org/10.3109/07388551.2011.595384

Pecharaply, A., Parkpian, P., Annachhatre, A. P., & Jugsujinda, A. (2007). Influence of anaerobic co-digestion of sewage and brewery sludges on biogas production and sludge quality. *Journal of Environmental Science and Health, Part A*, *42*(7), 911–923.

Polastri, P., Moreira, W. M., dos Santos, D. F., Janeiro, V., Vareschini, D. T., & Gimenes, M. L. (2024). Kinetic modeling and synergistic effects of anaerobic co-digestion of bovine ruminal waste and brewery spent grain. *Journal of Environmental Chemical Engineering, 12*(2), 111929.

Potdukhe, R. M., Sahu, N., Kapley, A., & Kumar, R. (2021). Co-digestion of waste activated sludge and agricultural straw waste for enhanced biogas production. *Bioresource Technology Reports, 15*, 100769–100769. https://doi.org/10.1016/j.biteb.2021.100769

Rajagopal, R., Mass´e, D. I., & Singh, G. (2013). A critical review on inhibition of anaerobic digestion process by excess ammonia. *Bioresource Technology, 143*, 632–641.

Rajagopalan, G., Subramanian, B., & Higgins, M. (2022). *Quality of biogas derived from wastewater solids and co-digested organic wastes: A characterization study* [WRF report]. The Water Research Foundation.

Rauch-Williams, T. (2023). *Characterization and contamination testing of source separated organic feedstocks and slurries for co-digestion at resource recovery facilities* [WRF project report]. The Water Research Foundation.

Ripoll, V. C. A-G., Perez, M., & Solera, R. (2020). Improvement of biomethane potential of sewage sludge anaerobic co-digestion by addition of "sherry-wine" distillery wastewater. *Journal of Cleaner Production, 251*, 119667.

Salminen, E., & Rintala, J. (2002). Anaerobic digestion of organic solid poultry slaughterhouse waste—A review. *Bioresource Technology*, *83*(1), 13–26.

Sawatdeenarunat, C., Surendra, K. C., Takara, D., Oechsner, H., & Khanal, S. K. (2015). Anaerobic digestion of lignocellulosic biomass: Challenges and opportunities. *Bioresource Technology, 178*, 178–186. http://dx.doi.org/10.1016/j.biortech.2014.09.103

Semberg, C., Macintosh, C., Astals, S., & Koch, K. (2019). Benefits and drawbacks of food and dairy waste co-digestion at a high organic loading rate: A Moosburg WWTP case study. *Waste Management*, *95*, 217–226. https://doi.org/10.1016/j.wasman.2019.06.008

Shrestha, S., Fonoll, X., Khanal, S. K., & Raskin L. (2017). Biological strategies for enhanced hydrolysis of lignocellulosic biomass during anaerobic digestion: Current status and future perspectives. *Bioresource Technology, 245*, 1245–1257. https://doi.org/10.1016/j.biortech.2017.08.089

Sosnowski, P., Wieczorek, A., & Ledakowicz, S. (2003). Anaerobic codigestion of sewage sludge and organic fraction of municipal solid wastes. *Advances in Environmental Research*, *7*, 609–616. https://doi.org/10.1016/S1093-0191(02)00049-7

Suto, P., Gray, D. M. D., Larsen, E., & Hake, J. (2006). Innovative anaerobic digestion investigation of fats, oils, and grease. *Proceedings of the Water Environment Federation, 22*(22), 858–879.

Szaja, A., & Montusiewicz, A. (2019). Enhancing the co-digestion efficiency of sewage sludge and cheese whey using brewery spent grain as an additional substrate. *Bioresource Technology, 291*, 121863.

Szaja, A., Montusiewicz, A., & Lebiocka, M. (2020). Anaerobic co-digestion of brewery spent grain and municipal sewage sludge under mesophilic and thermophilic conditions. In *The Role of Agriculture in Climate Change Mitigation* (pp. 129–135). CRC.

Szaja A., Montusiewicz, A., & Lebiocka, M. (2023). Variability of micro- and macro-elements in anaerobic co-digestion of municipal sewage sludge and food industrial by-products. *International Journal of Environmental Research and Public Health, 20*(7), 5405. https://doi.org/10.3390/ijerph20075405

Szaja, A., Montusiewicz, A., Lebiocka, M., & Bis, M. (2020). The effect of brewery spent grain application on biogas yields and kinetics in co-digestion with sewage sludge. *PeerJ*, 8:e10590. https://doi.org/10.7717/peerj.10590

U.S. Environmental Protection Agency. (2012, September). *Increasing anaerobic digester performance with codigestion*. https://www.epa.gov/sites/default/files/2014-12/documents/codigestion.pdf

U.S. Environmental Protection Agency. (2012). *Industrial food processing waste analyses*. Office of Resource Conservation and Recovery.

U.S. Environmental Protection Agency. (2001). *Method 1684—Total, fixed, and volatile solids in water, solids, and biosolids*. Office of Water Office of Science and Technology Engineering and Analysis Division (4303).

Wang, X., Yang, G., Feng, Y., Ren, G., & Han, X. (2012). Optimizing feeding composition and carbon–nitrogen ratios for improved methane yield during anaerobic co-digestion of dairy, chicken manure and wheat straw. *Bioresource Technology, 120*, 78–83.

Wu, Y.-R., & He, J. 2013. Characterization of anaerobic consortia coupled lignin depolymerization with biomethane generation. *Bioresource Technology, 139*, 5–12. https://doi.org/10.1016/j.biortech.2013.03.103

Xia, Y., Massé, D. I., McAllister, T. A., Kong, Y., Seviour, R., & Beaulieu, C. (2012). Identity and diversity of archaeal communities during anaerobic co-digestion of chicken feathers and other animal wastes. *Bioresource Technology, 110*, 111–119. https://doi.org/10.1016/j.biortech.2012.01.107

York, R. V., & Magner, J. A. (2009). Increasing CHP productivity while reducing biosolids volume and climate changing gasses. *Proceedings of the Water Environment Federation, 2009*(9), 6639–6651.

Zhang, W., Zhang, L., & Li, A. (2015). Anaerobic co-digestion of food waste with MSW incineration plant fresh leachate: Process performance and synergistic effects. *Chemical Engineering Journal, 259*, 795–805.

Zupančič, G. D., & Jemec, A. (2010). Anaerobic digestion of tannery waste: Semi-continuous and anaerobic sequencing batch reactor processes. *Bioresource Technology*, *101*(1), 26–33.

Zupančič, G. D., Uranjek-Ževart, N., & Roš, M. (2008). Full-scale anaerobic co-digestion of organic waste and municipal sludge. *Biomass and Bioenergy*, *32*(2), 162–167.

3.0 FURTHER READING

Ammenberg, J., & Roozbeh, F. (2017). Assessment of feedstocks for biogas production, part II—Results for strategic decision making. *Resources, Conservation and Recycling*, *122*, 388–404. https://doi.org/10.1016/j.resconrec.2017.01.020

Atamer, Z., Post, A. E., Schubert, T., Holder, A., Boom, R. M., & Hinrichs, J. (2017). Bovine β-casein: Isolation, properties and functionality. A review. *International Dairy Journal*, *66*, 115–125.

Feiz, R., & Ammenberg, J. (2017). Assessment of feedstocks for biogas production, part I—A multi-criteria approach. *Resources, Conservation and Recycling*, *122*, 373–387. https://doi.org/10.1016/j.resconrec.2017.01.019

Korniłłowicz-Kowalska, T., & Bohacz, J. (2011). Biodegradation of keratin waste: Theory and practical aspects. *Waste Management*, *31*(8), 1689–1701. https://doi.org/10.1016/j.wasman.2011.03.024

Menon, A., Wang, J.-Y., & Giannis, A. (2017). Optimization of micronutrient supplement for enhancing biogas production from food waste in two-phase thermophilic anaerobic digestion. *Waste Management*, *59*, 465–475.

Onifade, A. A., Al-Sane, N. A., Al-Musallam, A. A., & Al-Zarban, S. (1998). A review: Potentials for biotechnological applications of keratin-degrading microorganisms and their enzymes for nutritional improvement of feathers and other keratins as livestock feed resources. *Bioresource Technology*, *66*(1), 1–11. https://doi.org/10.1016/s0960-8524(98)00033-9

U.S. Environmental Protection Agency. (n.d.). *AgStar: Biogas recovery in the agricultural sector*. https://www.epa.gov/agstar

U.S. Environmental Protection Agency. (2014, September). *Food waste to energy: How six water resource recovery facilities are boosting biogas production and the bottom line*. https://www.epa.gov/sites/default/files/2016-07/documents/food_waste_to_energy_-_final.pdf

Chapter 4

Digestate, Dewatering, Sidestream, and Cake Quality Impacts

Matt Higgins

1.0 EFFECT OF CO-DIGESTION ON DIGESTATE AND BIOGAS CHARACTERISTICS

Addition of different feedstocks will impact digestate chemistry and characteristics depending on the characteristics of the feedstocks. The following sections consider, in more detail, the different impacts such as chemistry, gas composition, rheology, dewatering, and cake quality in terms of odors as well as microconstituents and how these parameters may impact the operation of the digester.

1.1 Using Stoichiometry to Understand Feedstock Impacts

The digestate and biogas chemistry and characteristics are greatly influenced by the feedstocks added to a digester. Stoichiometry is an excellent tool to help understand these impacts quantitatively and qualitatively. Stoichiometry is simply the application of mass balances at an elemental level. For example, if six moles of carbon are being utilized as reactants in a reaction, those six moles must also show up in the products according to conservation of mass principles.

A stoichiometry analysis is a steady state model; in other words, stoichiometry predicts what will happen when the reactants have transformed to their products. This approach does not consider the kinetics or rates of the reaction, which can be limiting in some cases of co-digestion, especially when first introducing a new co-substrate that readily degrades. If the loading rate is too high initially, an imbalance between the acid producers and methanogens could result in process instability. Stoichiometry would not predict this outcome and instead loading rates and kinetics would need to be considered to reduce the risk of process upset due to an imbalance in reaction rates. The stoichiometric analysis below assumes no imbalance in kinetics and is a steady state analysis.

As discussed previously, anaerobic digestion is a multi-step process in which different organisms are involved in the different steps such as hydrolysis, acidogenesis, acetogenesis, and methanogenesis (as discussed in Chapter 3). Despite being a multi-step reaction, the overall stoichiometry of the process can be utilized to understand the impacts of different feedstock characteristics. For example, Buswell and coworkers presented equations for the overall stoichiometry of anaerobic digestion with organic matter going to the final products of carbon dioxide and methane (Buswell & Boruff, 1932; Buswell & Mueller, 1952; Neave & Buswell, 1930). Based on this reaction, the carbon, hydrogen, and oxygen content of the feedstock would then impact the final products of methane and carbon dioxide.

This equation has been expanded to include other components in the feedstock such as nitrogen and sulfide. In addition, others have included the production of biomass because some of the elements go to production of microbial cells, although only about 5%–7% of the feedstock goes to formation of cells due to the low yields in anaerobic systems. For the purpose of our analysis, we will consider the elements of C, H, N, and O, along with the formation of new cells (with chemical composition $C_5H_7O_2N$), and also recognize that as ammonia is released from protein degradation, it will accept a hydrogen ion from carbonic acid to form ammonium and bicarbonate. This overall stoichiometric reaction would be (adapted from Rittman & McCarty, 2001):

$$C_aH_bO_cN_d + vH_2O \rightarrow wC_5H_7O_2N + xCH_4 + yCO_2 + zNH_4^+ + zHCO_3^- \quad (4.1)$$

The main concept of this stoichiometry is to consider that depending on the chemical makeup of the feedstock (the subscripts a, b, c, and d), this will impact the stoichiometric coefficients (w, x, y, and z) of the different products. By balancing the reactions and understanding how much of the feedstock degrades (as measured by volatile solids reduction [VSR]), we can estimate the biogas amounts and composition of methane and carbon dioxide (stoichiometric coefficients x and y, respectively), ammonium and bicarbonate concentration (stoichiometric coefficient z). In addition, because bicarbonate is the main type of alkalinity in a digester, we can estimate the effect of different feedstocks on digester alkalinity, and subsequent changes in pH because bicarbonate and CO_2 equilibrium can be used to estimate pH. The full calculations and equations of anaerobic digester stoichiometry are outside the scope of this text, but for additional information and a full description of these equations, the reader is referred to the literature cited in this section.

When considering the makeup of different feedstocks, we can generally consider the three main components of most organics: protein, carbohydrates, and fats. The typical elemental composition of carbohydrates, proteins, and fats is shown in Table 4.1. In addition, Higgins and Rajagopalan (2017) analyzed samples of different feedstocks as part of a co-digestion study, and the average elemental C, H, N, O compositions for these materials are also summarized in Table 4.1 along with results from analysis of primary and secondary solids from water resource recovery facilities (WRRFs), and these are compared to several literature values. For normalizations, when compounds contain N, the stoichiometric subscript for N were all set to 1 and the other elemental subscripts were adjusted accordingly. This allows an easier comparison of the C/N ratio the implications of which will be discussed in more detail in subsequent sections.

Using the stoichiometric composition of a particular feedstock along with Equation 4.1 and an estimate of the volatile solids reduction, the impacts of a particular feedstock on parameters such as pH, alkalinity, biogas production and composition, and ammonia concentration can be estimated, which is a very powerful tool for understanding feedstock impacts.

As a note, the sulfur content of the feedstock could also be included in the stoichiometry, although sulfur chemistry in a digester is quite complex and difficult due to pH, complexation, and precipitation reactions that can occur.

A series of stoichiometric model runs was performed to demonstrate the impacts of the different components (fat, carbohydrate, protein) of potential

TABLE 4.1 Summary of Elemental Compositions of Different Feedstocks

Feedstock	Elemental Composition	C/N Ratio	Source
Typical Carbohydrate	$C_6H_{10}O_5$	High	Rittmann & McCarty, 2001
Typical Protein	$C_4H_6O_{1.3}N$	4	Rittmann & McCarty, 2001
Typical Fat	$C_{16}H_{32}O_2$	High	Rittmann & McCarty, 2001
Average Food Waste	$C_{17}H_{30}O_{6.4}N$	17	Higgins & Rajagopalan, 2017
Average FOG	$C_{131}H_{258}O_{20}N$	131	Higgins & Rajagopalan, 2017
Avg. Primary Solids	$C_{17}H_{31}O_{7.2}N$	18	Higgins & Rajagopalan, 2017
Primary Solids	$C_{18}H_{36}O_{10}N$	28	Sotemann et al., 2005
Avg. Secondary Solids	$C_{6.6}H_{11.8}O_{2.4}N$	6.6	Higgins & Rajagopalan, 2017
Secondary Solids	$C_{7.4}H_{14}O_4N$	7.4	Ekama, 2009

co-substrates. To make an apples-to-apples comparison, each co-substrate feed concentration was assumed to be 10% with no inert material and each had a VSR of 85%. Table 4.2 summarizes the inputs that were chosen for the stoichiometric model in terms of the feedstock and solids characteristics. The model assumed the same mass of primary/secondary solids was added to the digester, and then additional amounts of the co-substrates were added to achieve an increase in volatile solids (VS) loading to the digester of 10%, 25%, 50%, and 75%. The indigenous primary and secondary solids from the WRRF were fed at a constant VS loading rate for all scenarios, which was 2.13 kg VS/m^3-d with an SRT of 20 days. Based on the values from Table 4.2, the loading rates and SRT for each scenario were calculated and are summarized in Table 4.3. The SRT decreases as more co-substrate is added to the digester, but in the scenarios described, the SRT remains above 15 days which would likely not impact overall digestion such as the VSR.

TABLE 4.2 Summary of Stoichiometric Inputs for Comparing Co-digestion of Different Co-substrates

Feedstock	Elemental Composition	Feed TS (%)	Feed VS/TS Ratio	VSR of Individual Component (%)
Primary/Secondary Blend	$C_{9.8}H_{17.8}O_{4.2}N$	5	0.85	55
Carbohydrate	$C_6H_{10}O_5$	10	1.0	85
Protein	$C_4H_6O_{1.3}N$	10	1.0	85
Fat	$C_{16}H_{32}O_2$	10	1.0	85

TABLE 4.3 Summary of Volatile Solids Loading and Digester Solids Retention Time for the Varying Scenarios

Feedstock	VS Loading Rate (kg VS/m^3-d)	Digester SRT (d)
Primary/Secondary Blend	2.13	20.00
10% Additional VS from Co-substrate	2.34	19.18
25% Additional VS from Co-substrate	2.66	18.08
50% Additional VS from Co-substrate	3.19	16.49
75% Additional VS from Co-substrate	3.72	15.17

The results for the different parameters based on the model are summarized in the following sections.

1.1.1 pH and Alkalinity

The impact of different feedstocks on digester pH and alkalinity can be understood in part by stoichiometry and estimated through the equilibrium of bicarbonate and dissolved carbon dioxide. A full treatment of these reactions and calculations is outside the scope of this text, but for additional information, the reader is referred to the literature (Ekama, 2009; Rittman & McCarty, 2001; Sotemann et al., 2005). Other reactions could also occur within the digester such as precipitation and metal hydrolysis; however, the main drivers for changes in pH can be estimated from the stoichiometry. To demonstrate these potential impacts, a stoichiometric analysis was performed with a typical primary/secondary blend and then co-substrate addition of either carbohydrates, fats, and proteins at different loading rates as described in the previous section (see Table 4.2).

The results from this analysis for pH and alkalinity are summarized in Figures 4.1 and 4.2, respectively. For the case of fats and carbohydrates, their addition results in a decrease in digester pH and alkalinity, while the addition of protein increases both. The difference in the pH and alkalinity response can be explained by the nitrogen content of each co-substrate. As protein is degraded, the nitrogen is released as ammonia, which is a base and accepts a proton donated from carbonic acid to form bicarbonate. The result is an increase in pH as well as alkalinity. Referring back to Equation 4.1, the stoichiometry shows that for each mole of ammonia formed, one mole of bicarbonate will also be formed. Organics that do not contain nitrogen will result in a net decrease in pH and alkalinity, while nitrogen containing organics will tend to raise pH and alkalinity. This is an important

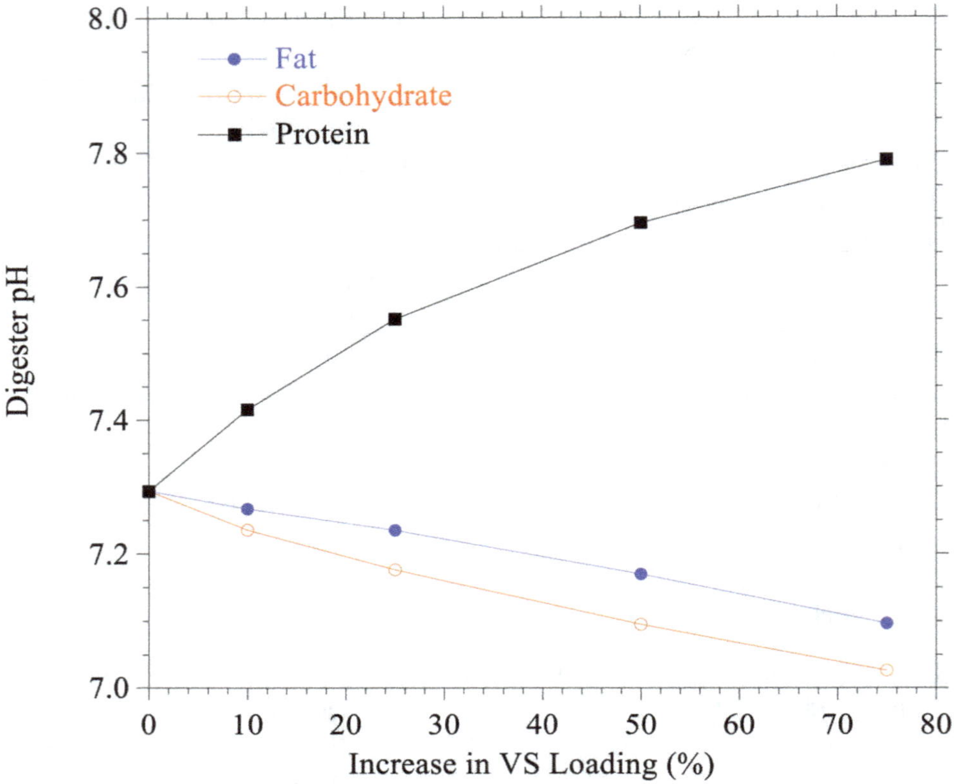

FIGURE 4.1 Effect of Co-substrate Addition on Digester pH

consideration when evaluating different feedstocks because maintaining stable digester operation is dependent on maintaining pH and alkalinity; therefore, a balance in the feedstock formulation would be preferable.

1.1.2 Digester Total Ammonia Nitrogen and Phosphorus Concentration

As discussed in the previous section, the production of ammonium is equivalent to the production of bicarbonate that arises from degradation of organic nitrogen during anaerobic digestion. The concentration of ammonium in the digester as a function of co-substrate loading is shown in Figure 4.3. Similar to alkalinity, the ammonium concentration decreases with the addition of fat or carbohydrates, but increases with the addition of protein. One concern of a high-protein feedstock would be ammonia inhibition, which can occur at ammonium concentrations greater than about 2500 mg NH_4^+-N/L. As shown in Figure 4.3, this can happen with a relatively low loading of a highly proteinaceous feed.

Most co-substrates do not have significant concentrations of phosphorus that contribute appreciably to the phosphorus concentrations in digesters,

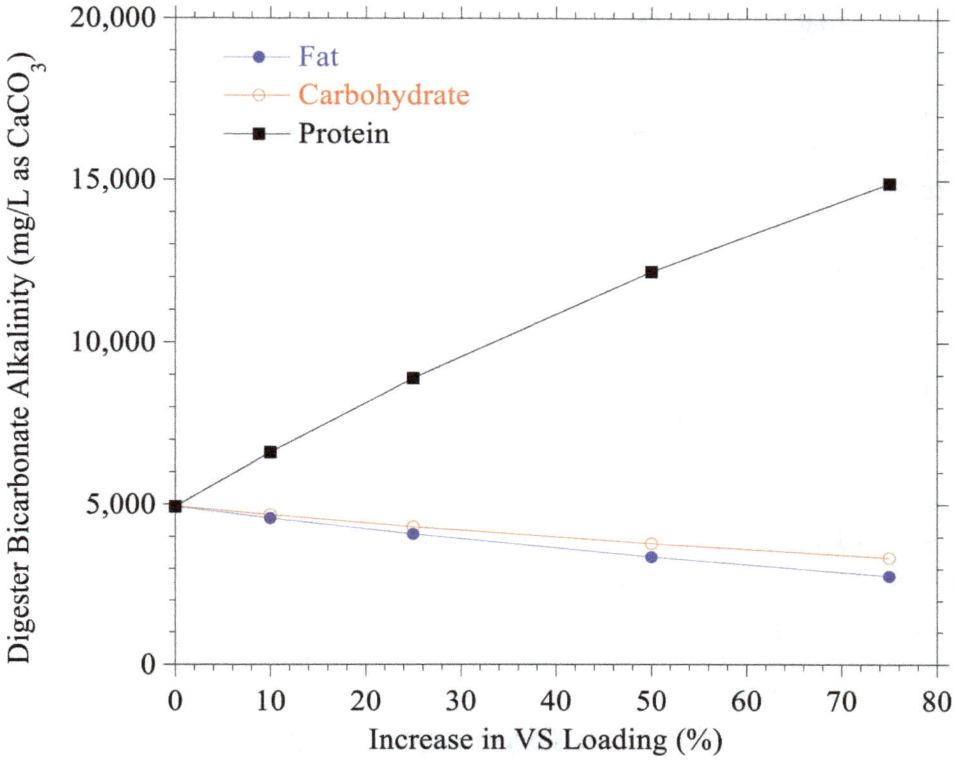

FIGURE 4.2 Effect of Co-substrate Addition on Digester Bicarbonate Alkalinity

but testing of feedstocks for phosphorus should be performed to assess this possibility, especially when nutrient limitations in the cake and recycle are a concern. The topic of ammonium and phosphate in the recycle streams is discussed further in Section 2.3 including results from several trials in which phosphate was measured for co-digestion of different wastes.

1.1.4 Total Solids/Volatile Solids

As more co-substrates are fed to the digester, the digester effluent total solids (TS) and VS concentrations could change. Typically, the co-substrates have higher TS and VS values than the indigenous solids from the WRRF and because they do not completely degrade, the effluent TS and VS concentrations could increase but not always. In the stoichiometry scenarios for this chapter, the TS and VS of each co-substrate (fat, carbohydrate, protein) were all the same with the same VSR; therefore, the effluent TS and VS concentrations will be the same. The effect on solids as well as the mass of dry solids leaving the facility is summarized in Table 4.4.

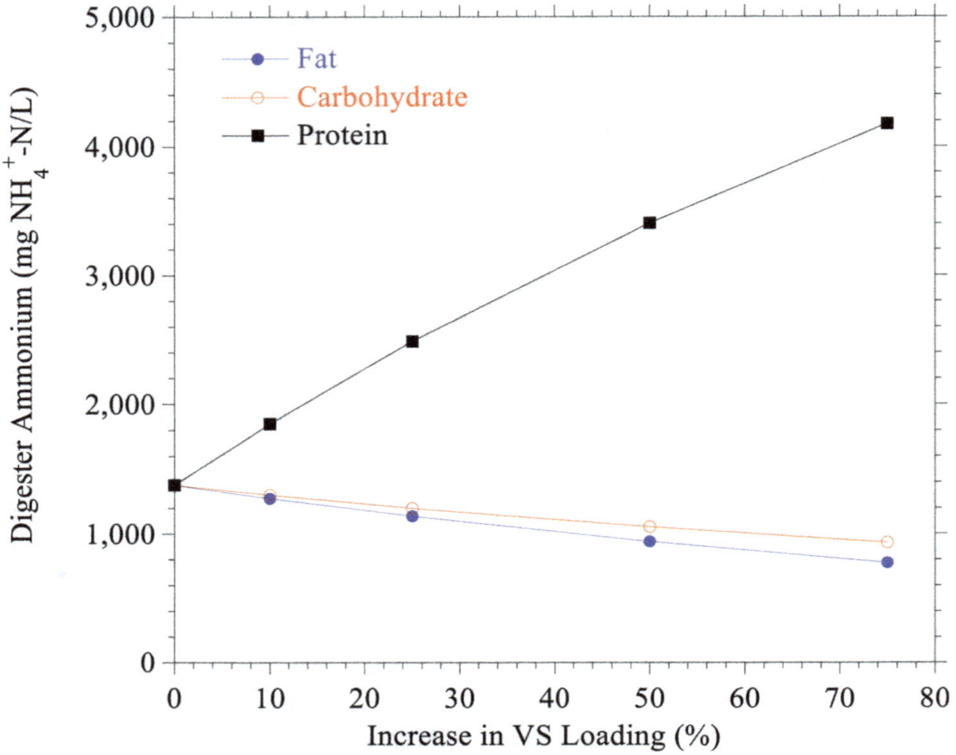

FIGURE 4.3 Effect of Co-substrate Addition on Digester Ammonium Concentration

TABLE 4.4 Summary of Effluent Solids Characteristics for Stoichiometry Analysis

Feedstock	Effluent TS (%)	Effluent VS (%)	Effluent VS/TS	Change in Dry Mass Leaving WRRF (%)
Primary/Secondary	2.66	1.91	0.72	0.00
10% Additional VS from Co-substrate	2.62	1.90	0.72	2.39
25% Additional VS from Co-substrate	2.55	1.87	0.73	5.99
50% Additional VS from Co-substrate	2.46	1.84	0.75	12.0
75% Additional VS from Co-substrate	2.38	1.81	0.76	18.0

It is interesting to note that effluent TS and VS actually decrease with the additional co-substrate addition, and this is due to better degradation of the co-substrate, which lowers their effluent values below the primary/secondary blend resulting in a dilution of the TS and VS in the digester. However, the mass of dry solids leaving the facility does increase relative to the control scenario (only primary/secondary solids) because the co-substrates do not degrade completely, and the residual solids will increase the mass of dry solids leaving the facility. When considering the dry mass yields, addition of 75% additional VS to the digester only results in about 18% addition dry total solids exiting the digester for this scenario that was modeled. Whether TS and VS increase in the digester will be a function of the feed concentration of the co-substrate along with the VSR of that co-substrate, and in many cases, the TS and VS do increase for co-digestion of high concentration wastes.

The wet solids leaving the facility will depend on any impacts the co-substrates have on dewatering, specifically the cake solids after dewatering and this is discussed in more detail in the dewatering section (Section 2.0).

1.1.5 Biogas Production and Composition

The effect of different feedstock characteristics on production of biogas and methane can be evaluated with several different parameters using the stoichiometric analysis. The biogas and methane yields per mass of the individual feedstock provides insights into the differences between feedstocks. Table 4.5 summarizes the biogas and methane yields in terms of kilograms of VS fed to the digester, while Figure 4.4 shows the biogas and methane production volumes as function of the increase in VS loading from the three components (protein, fats, carbohydrates).

For comparison, Figure 4.5 shows the increase in methane production from different full-scale and laboratory scale data sets. The increase in methane production is fairly consistent between the different studies as they are

TABLE 4.5 Summary of Feedstock on Biogas and Methane Yields

Feedstock	Methane Content of Biogas (%)	Methane Yield (L CH_4/kg VS_{fed})	Biogas Yield (L Biogas/kg VS_{fed})
Primary/Secondary Blend	65	350	540
Carbohydrate	49	370	750
Protein	69	470	680
Fat	72	890	1240

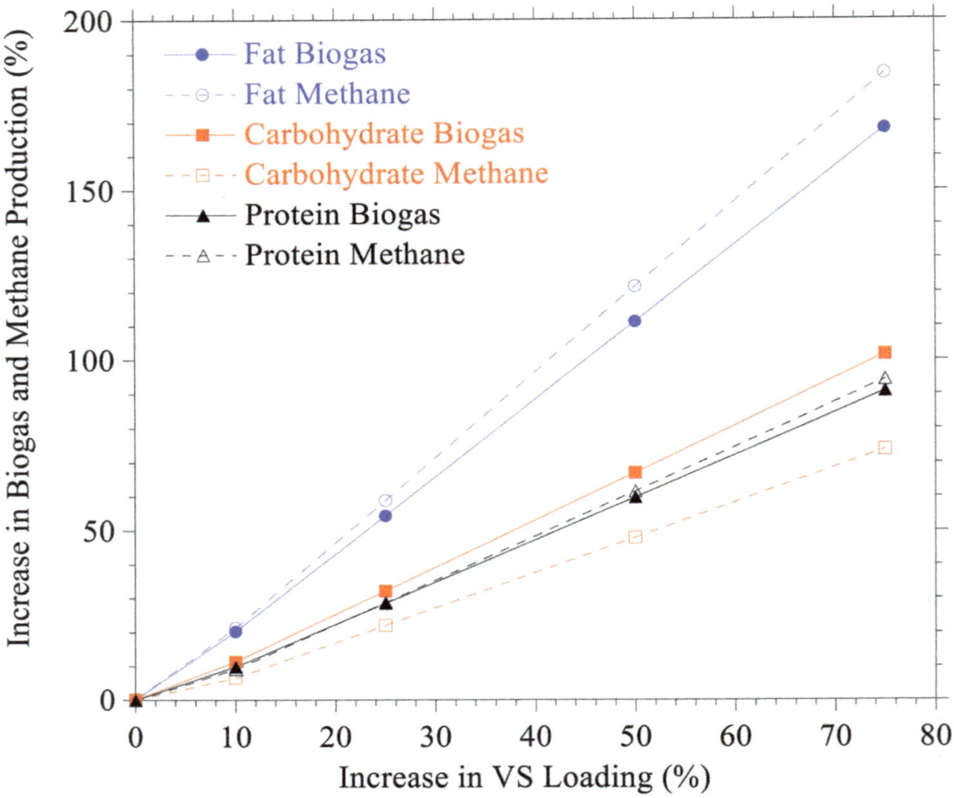

FIGURE 4.4 Effect of Co-substrate Addition on Biogas Methane Production Based on Stoichiometry

all food wastes that are a mix of proteins, fats, and carbohydrates. The average slope of the line is approximately 1.5, which means that for the addition of an x defined percent increase in a co-substrate, this will result in a $1.5x\%$ increase in the methane production.

Fats produce the highest amount of methane per mass of VS fed to the digester due to their high energy content, while carbohydrates produce the least. The methane content of carbohydrates is relatively low, around 49% for the model, while about 72% for fats and 69% for protein. As a result, the methane content of the biogas will tend to decrease with the addition of a carbohydrate co-substrate.

In a Water Research Foundation–funded study, Rajagopalan et al. (2022) investigated the potential for a large number of different organic compounds (OCs) and other species to accumulate in the biogas. They reported that several organic wastes increased the levels of some OCs, especially acetone, butanone, ethanol, and butanol, while addition of co-substrates often reduced the concentration of siloxanes, likely due to dilution of the biogas with low siloxane concentration biogas from co-substrates.

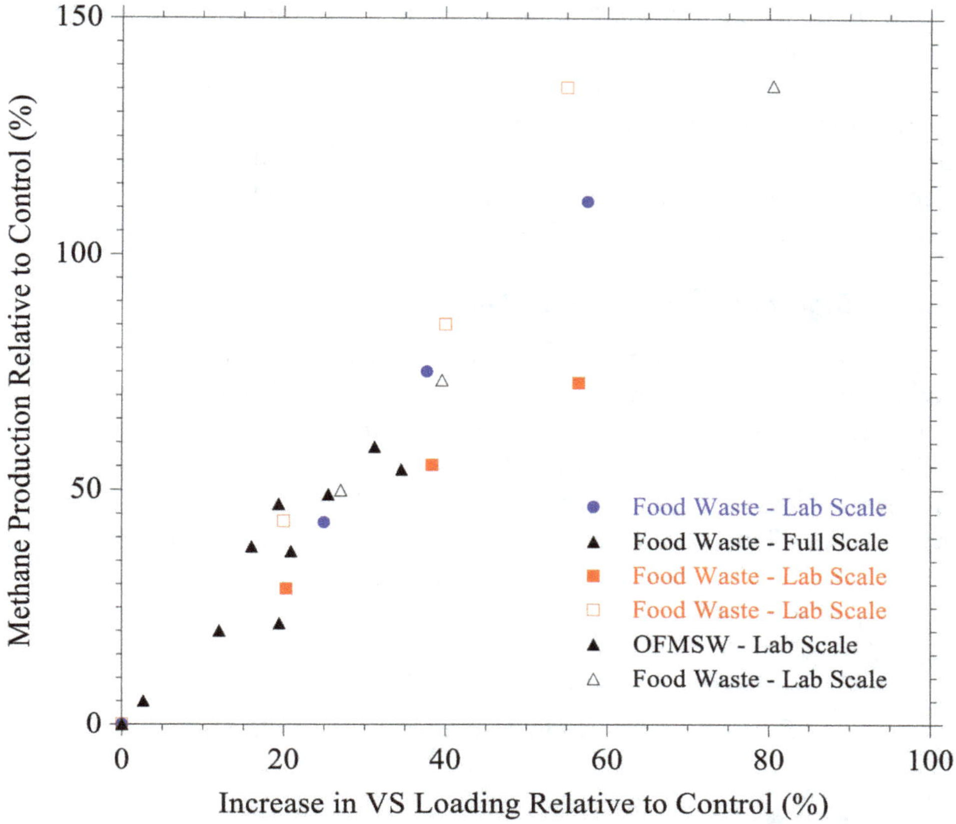

FIGURE 4.5 Effect of Volatile Solids Loading on Methane Production Increase Relative to a Control for Different Lab and Full-Scale Digester Studies

In addition, they reported that ammonia concentration in the biogas increased as the nitrogen content of the co-substrate increased (Rajagopalan et al., 2022). The increase in biogas ammonia concentration could be an issue as it could contribute to corrosion and increase NOx emission during biogas combustion (Chambers et al., 1996; Hasegawa, 2010). Higgins et al. (2019) also showed that as the digestate concentration of ammonium increased, the biogas ammonia concentration also increased. This would be expected from gas/liquid equilibrium between ammonia in the digestate and the ammonia in the biogas according to Henry's Law.

As discussed earlier, the addition of high nitrogen wastes such as protein will increase both the ammonium and pH, resulting in an increase in the biogas ammonia concentration. Unpublished research has shown a good correlation between the predicted biogas ammonia and the actual measured biogas ammonia based on equilibrium modeling, although the biogas ammonia was typically lower than the predicted value with measured values of biogas ammonia ranging from 20–260 ppm (Higgins, 2024).

Although sulfur was not included in the stoichiometric analysis due to the complexity of reactions that sulfur can participate in, one consideration is that high protein wastes could result in greater concentrations of reduced sulfur compounds in the biogas because several amino acids contain sulfur and their degradation leads to production of compounds such as hydrogen sulfide, methyl mercaptan, and dimethyl sulfide, which are all odorous compounds (Adams et al., 2007).

1.2 Digestate Rheology and Mixing

Digestate rheology is a critical parameter for the operation of a digester, impacting processes such as mixing, pumping, and dewatering. When the digestate solids concentration is above about 1.5%, the slurry has shear thinning behavior, meaning that the viscosity decreases as the shear rate increases. The shear rate can be thought of as the mixing intensity or velocity gradient. Figure 4.6 shows the digestate rheology for a digester that had

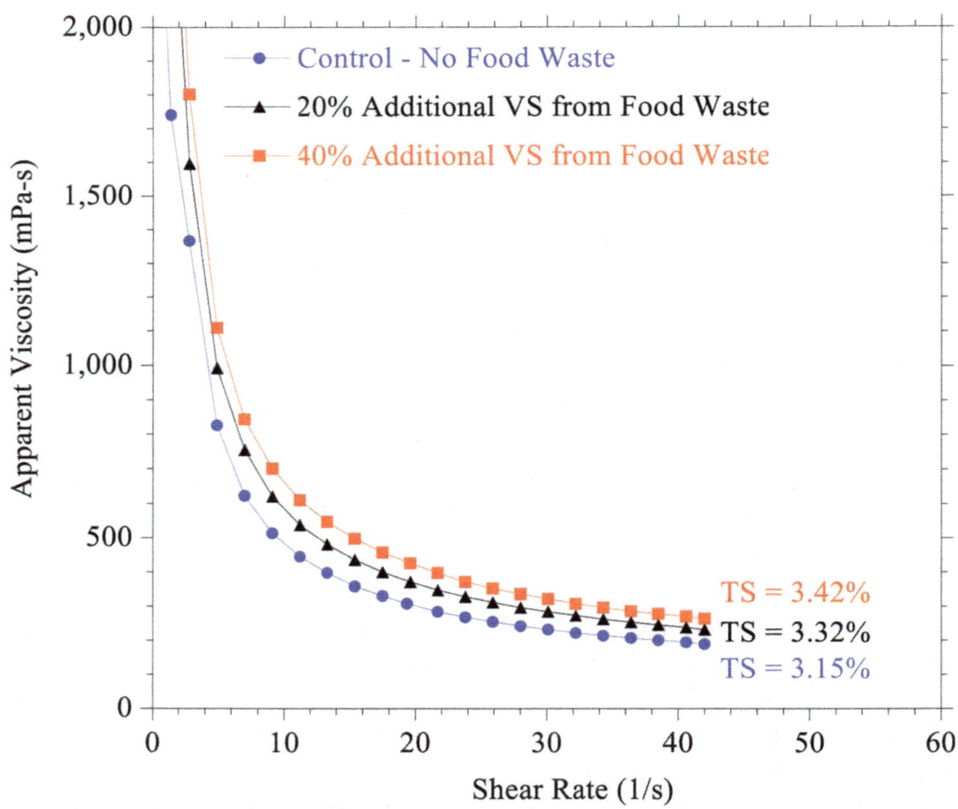

FIGURE 4.6 Apparent Viscosity Versus Shear Rate for a Control Digester With No Food Waste and Two Digesters With Increasing Food Waste Loading

a TS concentration of about 3% (Higgins, unpublished data). As the shear rate increases (plotted on the x-axis), the apparent viscosity decreases.

The term "apparent" is often used to qualify viscosity because the viscosity is not a single number, but varies depending on shear rate.

One of the factors that has a significant impact on digestate rheology is solids concentrations. As the TS concentrations increase, the viscosity will also typically increase. Figure 4.6 shows curves from a laboratory scale trial in which a control digester was fed solids from the facility, and then two digesters received the same indigenous solids along with increasing loadings of source separated food waste—one digester received about 20% additional VS while the other received 40% additional VS. As a result of the increased loading to the digester, the TS concentrations increased at the higher loading rates by about 5% for the 20% loading and about 9% for the 40% additional VS loading digester. The apparent viscosity curves increased as the TS concentration increased. The increase in viscosity experienced in the digester will depend on the shear rate of mixing in the digester, but within the range of shear rates tested, the viscosities are all greater than the control mainly due to the high TS of the feed in this case. The TS will not always increase with co-digestion as discussed previously (see Table 4.4) and will depend on the solids concentration of the feeds and their respective VSRs.

The results show that co-digestion could potentially impact mixing and viscosity in the digester, which is an important consideration because these are related to foaming and volume expansion potential in a digester (Chapman & Krugel, 2011; Higgins et al., 2014).

2.0 DEWATERING AND SIDESTREAM IMPACTS

One of the key drivers in solids handling economics is the dewatering and hauling costs. As more organics were loaded into a digester, this could increase the solids coming out of the digester, especially the dewatered cake production if the dewatering is impacting negatively. A significant amount of work has been done to evaluate the impacts of co-digestion on dewatering, especially the effect on polymer demand and cake solids. The following sections review some aspects of these two topics.

2.1 Effect on Polymer Demand

To condition solids before dewatering, most WRRFs use high molecular weight, cationic polymer. These polymers are positively charged, which interact with the negatively charged particles in the digestate to aggregate them into flocs that can withstand the shear of mechanical dewatering. One of the key factors that impacts polymer demand is the concentration of biocolloids in solution (Higgins, Murphy, & Chen, 2006; Murthy & Novak, 1999; Murthy et al., 2000, Novak et al., 2004). These biocolloids comprise mainly protein and some polysaccharides and research has shown that as the concentration of these particles increase, the polymer demand increases proportionally.

The addition of co-substrate could increase the polymer dose (in terms of kilograms of active polymer per ton of dry solids) if more biocolloids are created, but also even if the polymer dose is the same, the greater amounts of solids exiting the digester would increase the polymer demand.

Figure 4.7 summarizes data compiled from different full-scale and laboratory scale co-digestion studies. The *x*-axis shows the percent increase in the VS loading to the digester relative to the control scenario only being fed the indigenous solids, and the *y*-axis shows the relative change in polymer dose compared to the control. The same polymer was used for all the testing, which was a high molecular weight, linear cationic polymer. The horizontal dotted line shows where no impact would be observed relative to the control. The graph shows that for many of the projects, the polymer dose increased,

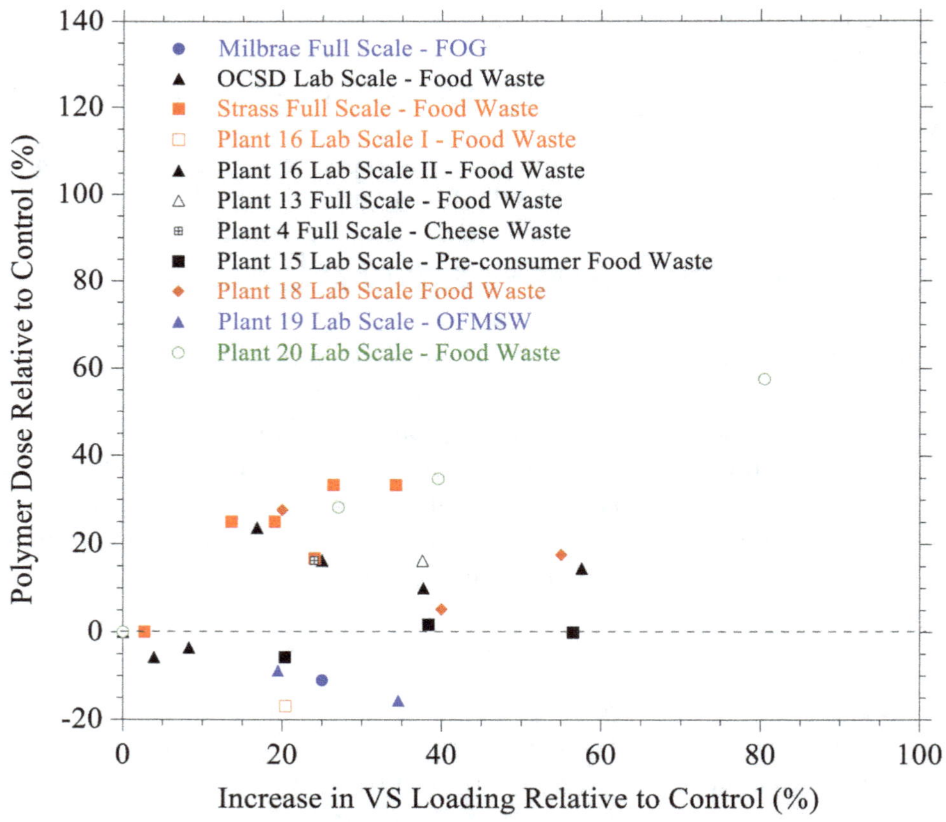

FIGURE 4.7 Polymer Dose Relative to a Control as a Function of Increased Volatile Solids Loading From Co-substrates

especially as loading rates increased, although in several cases the polymer dose actually decreased. As a note, all of the laboratory scale trials were performed at one university that used the same digestion and laboratory scale dewatering protocol—for a description of the protocol, see Higgins et al. (2017). The potential for increased polymer demand could impact the economics of co-digestion due to the relatively high costs of polymer, and should be a consideration in the planning stages of evaluating feedstocks.

2.2 Effect on Cake Solids and Wet Solids Production

Similar to impacts on polymer demand, the impacts of co-digestion on cake solids can have a substantial effect on the economics of co-digestion. If cake solids decrease, this could lead to greater hauling costs. The same data set for polymer demand was used to estimate the dewatered cake solids production for the different VS loadings from co-substrates for studies in which this data was available. The results are shown in Figure 4.8. The dewatered

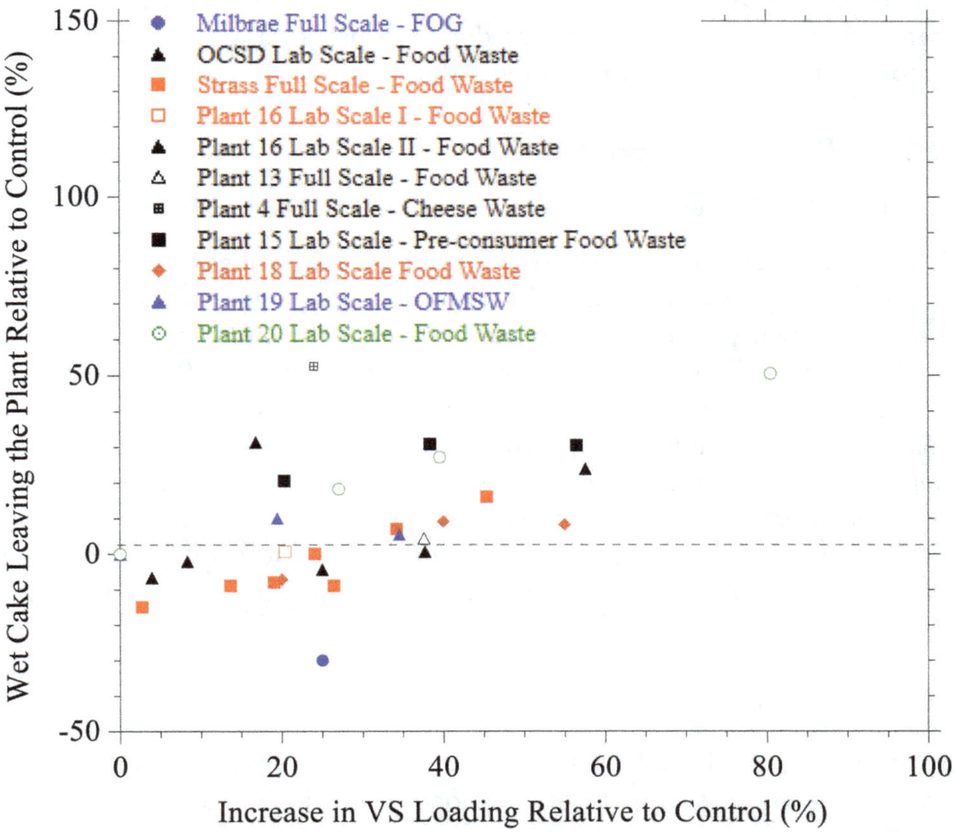

FIGURE 4.8 Dewatered Cake Production to a Control as a Function of Increased Volatile Solids Loading From Co-substrates

cake production includes the impacts of VSR as well as the dewaterability in terms of cake solids. Interestingly, the dewatered cake leaving the facility was often less than the control for lower loading rates, less than about 20%, but as the loading rates increase, the dewatered cake production tends to increase above the control scenario. The lower mass of dewatered cake at lower loading rates is due to higher cake solids that can occur for some wastes. Dewatering is impacted by numerous factors including the C/N ratio and the cation balance and co-digestion could impact these factors. In addition, researchers have speculated that co-digestion can create a synergistic effect through stimulation of the biomass which leads to additional VS destruction of the indigenous solids (Aichinger et al., 2015; Kim et al., 2017; Navarantnam et al., 2011; Zitomer et al., 2008).

As the loadings increase, eventually the greater mass of solids fed to the digester will increase the wet mass production, which is shown by the data in Figure 4.8. The effect on cake solids and dewatered cake production is an important consideration in any co-digestion plan because it can have a large effect on the economics of the solids management process. For example, Higgins et al. (2017), showed that hauling costs are a critical factor in determining the economics of a co-digestion program and could limit the amount of co-substrates to be imported.

2.3 Effects on Sidestreams

2.3.1 Solids Capture

The effect of co-digestion on solids capture has not been as well documented as polymer demand and cake solids, but in general, co-digestion has not shown positive or negative trends on the solids capture. For a well-conditioned digestate, the capture is typically good, >95%. As noted in the previous section, polymer demand often increases with co-digestion loading and this additional polymer is often required for creating good solids capture. A recent laboratory scale study showed that for three different loadings of food waste, the capture of solids was greater than 98%. Additional research is needed to further document effects on solids capture so the industry can more fully understand the potential implications of co-digestion on recycle streams.

2.3.2 Nutrients, N and P

Co-digestion could result in an increase in the nutrient loads from recycle streams. In the case of nitrogen, this load would be due to an increase in ammonia from the biodegradation of organic nitrogen mainly associated with protein. Referring back to Figure 4.3, addition of a high protein waste can greatly increase the ammonia concentration that would end up in the recycle stream. In comparison, a high fat or carbohydrate waste would tend to decrease the ammonia concentration in the recycle stream. Predicting the change in ammonia is straightforward if the nitrogen content of the co-substrates is known. The return of high concentrations of ammonia would increase treatment costs if nutrient removal is required; for example, aeration costs could increase and as well as potential costs for external carbon.

The phosphorus concentration for many co-substrates does not increase the digester concentrations unless the waste is from a specialty process that uses phosphate detergents or has other potential but unusual sources of phosphorus. Figure 4.9 shows data from three lab scale studies that fed food waste to the digester and in all cases, the phosphate concentrations remained the same or decreased relative to the control digester.

Nutrient loads could have economic impacts if nutrient removal is required on the liquid side of the treatment process. For example, Kang

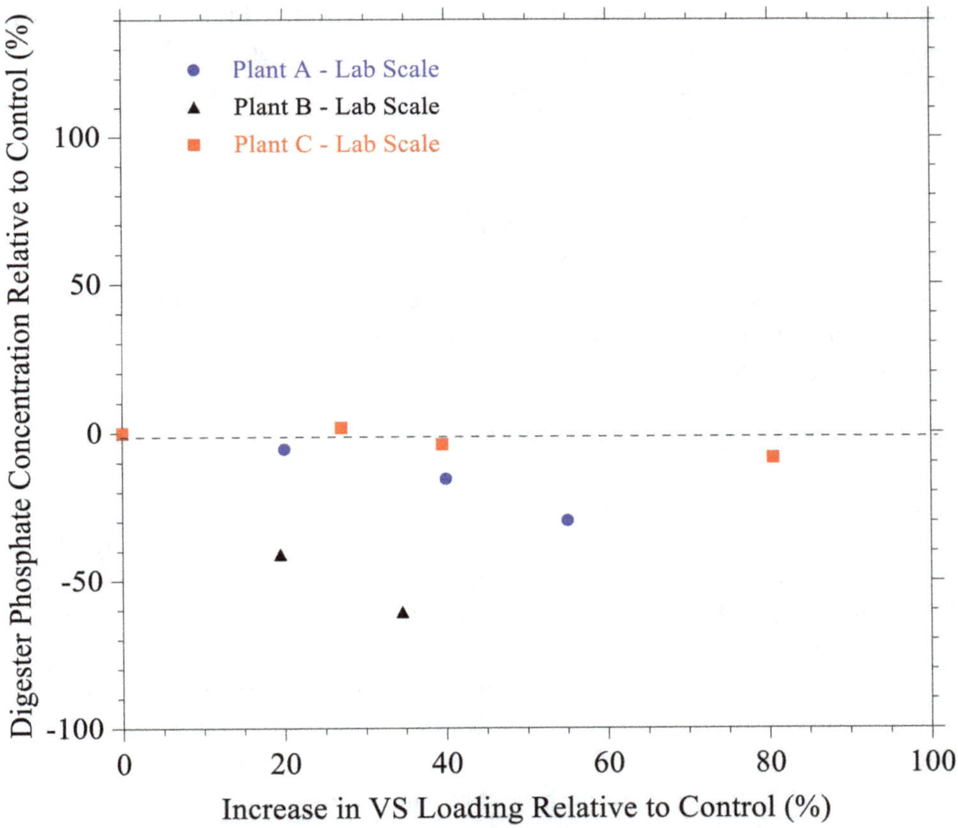

FIGURE 4.9 Digester Phosphate Concentration Relative to a Control as a Function of Increased Volatile Solids Loading From Co-substrates

et al. (2008) estimated unit costs for total phosphorus and total nitrogen removal in the range of 2–13 $/kg. As a result, increased nutrient loading should be considered as part of an economic model in evaluating co-digestion programs.

2.3.3 Struvite Formation Potential

Scaling due to struvite ($MgNH_4PO_4$-$6H_2O$) is a nuisance problem at many utilities that have relatively high concentrations of phosphate in the digestate, such as WRRFs with biological nutrient removal. Although co-digestion has not typically been reported to increase phosphate concentrations, when the co-substrates are high in protein, two factors could increase the potential for struvite formation. First, the degradation of protein by fermenters will release ammonium into solution which is one of the components of struvite, thereby increasing the struvite formation potential. Second, as discussed previously, protein degradation and the release of ammonia will increase the pH of the digester, which creates more favorable chemistry for the formation of struvite. As a result, WRRFs should consider the increased potential for struvite formation, especially in the case of high protein co-substrates as this could increase maintenance and costs associated with co-digestion.

3.0 CAKE QUALITY IMPACTS

Addition of co-substrates to anaerobic digester could impact the final product quality in terms of odorants or contaminants. The following sections review these two potential impacts.

3.1 Odorant Production

The production of nuisance odors is a significant concern for the viability of land application programs. Odors produced during cake storage have been studied in detail and volatile organic sulfur compounds (VOSCs) have been identified as key odorants produced during cake storage (Glindemann et al., 2006; Higgins, Yarosz et al., 2006; Novak et al., 2006). As a result, researchers have used the measurement of VOSCs as an indicator for the production of odors that could be released when cake is land applied. The three main VOSCs often evaluated are methyl mercaptan, dimethyl sulfide, and dimethyl disulfide. Adding these three together provides the total VOSCs or TVOSC as the indicator of odor potential from the cake.

The production of TVOSCs is largely a function of bioavailable materials, especially protein, in the cake, and the methanogenic activity (Higgins,

Yarosz et al., 2006; Higgins et al., 2008). In addition, the dewatering process can contribute to the release of bioavailable substrate which can fuel odorant production. For example, centrifuge dewatering that imparts significant shear to the solids has been shown to produce cakes with higher odorant production than lower shear devices such as belt filter presses.

Figure 4.10 shows the results from several laboratory scale co-digestion studies in which the TVOSC concentrations were measured during cake storage. In most cases, the odorant potential was less with the addition of co-substrates, at least for the TVOSC concentrations. This is likely due to the higher methanogenic activity in the cake because methanogens have been shown to be key organisms responsible for the degradation of VOSCs. The results suggest that co-substrates should have minimal impacts on cake odorant production as long as the co-substrates are well digested and in many cases the odorant production could be less than the control.

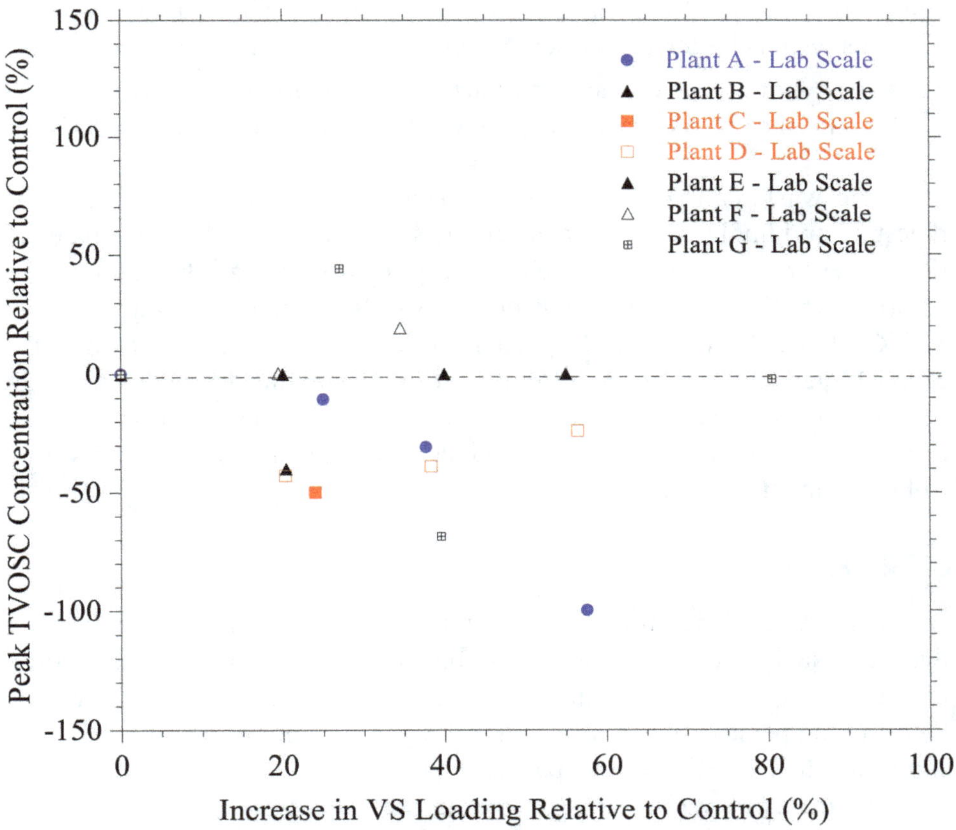

FIGURE 4.10 Peak TVOSC Concentrations Produced During Cake Storage Relative to a Control as a Function of Increased Volatile Solids Loading From Co-substrates

3.2 Microconstituents

Co-digestion of different wastes such as food waste or the organic fraction of municipal solid waste could affect the cake quality in terms of contaminants such as microplastics as well as chemicals such as per- and polyfluoroalkl substances (PFAS) used in packaging and pesticides/herbicides used in food production.

3.2.1 Microplastics

The U.S. Environmental Protection Agency (U.S. EPA) recently published a review of microplastics in food waste (U.S. EPA, 2021a) and provided the following conclusions:

- Food packaging and containers were the main source of plastic contamination of food waste, and the food waste itself also contributed to the plastic contamination.
- Prevention of food waste contamination will require public education as well as better source separation and reduction of contamination.
- Information and characterization of plastics in food waste is limited and in need of further research to better understand and address this issue.

Food waste can contribute to the overall load of microplastics in the digestate and final cake product after digestion. For example, Golwala et al. (2021) reported that food waste from a grocery store had 300 000 microplastic particles per kg of dried food waste. In comparison, sludges from WRRFs have a wide range of particle concentrations, ranging from 800–270 000 particles per kg. Screening of food waste and other co-substrates for microplastics would be beneficial to understanding the potential for changes in the plastics in the final product. This is an area of needed research to better understand the impacts.

3.2.2 Persistent Chemicals

Another area of concern is the potential to increase persistent chemicals in digestate, such as PFAS, pesticides, herbicides, or other products that could potentially be found in co-substrates. The U.S. EPA (2021b) reviewed the literature on persistent chemicals in food waste and reported that food waste can be a source of PFAS due to packaging with these chemicals, although the concentrations were generally lower than that found in biosolids. Similarly, food waste does not appear to be a significant source of pesticides. In general, screening of co-substrates for persistent chemicals would be suggested when there is concern of these chemicals impacting the final product quality.

4.0 REFERENCES

Adams, G. M., Witherspoon, J. R., Erdal, Z. K., Forbes, R. H., Hargreaves, J. R., Higgins, M. J., McEwen, D. W., & Novak, J. T. (2007). *Identifying and controlling the municipal wastewater odor environment phase 3: biosolids processing modifications for cake odor reduction* (Report No.03-CTS-9T). Water Environment Research Foundation.

Aichinger, P., Wadhawan, T., Kuprian, M., Higgins, M., Ebner, C., Fimml, C., Murthy, S., & Wett, B. (2015). Synergistic co-digestion of solid-organic-waste and municipal-sewage-sludge: 1 plus 1 equals more than 2 in terms of biogas production and solids reduction. *Water Research*, *87*, 416–423.

Buswell, A. M., & Boruff C. S. (1932). The relation between the chemical composition of organic matter and the quality and quantity of gas produced during sludge digestion. *Sewage Works Journal*, *4*, 454–460.

Buswell, A. M., & Muller, H. F. (1952). Mechanism of methane fermentation. *Industrial and Engineering Chemistry*, *44*, 550–552.

Chambers, A., Yoshii, Y., Inada, T., & Miyamoto, T. (1996). Ammonia decomposition in coal gasification atmospheres. *Canadian Journal of Chemical Engineering*, *74*, 929–934.

Chapman, T., & Krugel, S. (2011). Rapid volume expansion—an investigation into digester overflows and safety. *Proceedings of the Water Environment Federation*, USA, *4*, 1016–1040.

Ekama, G. A. (2009). Using bioprocess stoichiometry to build a plant-wide mass balance based steady-state WWTP model. *Water Research, 43*, 2101–2130.

Glindemann, D., Murthy, S. N., Higgins, M. J., Chen, Y. C., & Novak, J. T. (2006). Biosolids incubation method for odorous gas measurement from dewatered sludge cakes. *Journal of Residuals Science and Technology*, *3*, 153–160.

Golwala, H., Zhang, X., Md Iskander, S., & Smith, A. (2021). Solid waste: An overlooked source of microplastics to the environment. *Science of the Total Environment*, *769*.

Hasegawa, T. (2010). Gas turbine combustion and ammonia removal technology of gasified fuels. *Energies*, *3*, 335–449.

Higgins, M. J., Adams, G., Card, T., Chen, Y. C., Erdal, Z., Forbes Jr., R. H., Hargreaves, J. R., McEwen, D., Murthy, S. N., Novak, J. T., & Witherspoon, J. (2008). Role of protein, amino acids, and enzyme activity in odor production from anaerobically digested and dewatered biosolids. *Water Environment Research*, *80*, 127–135.

Higgins, M. J., Beightol, S., Nielsen, P. J., & Murthy, S. N. (2014). *Insights into rapid volume expansion in anaerobic digesters*. Proceedings of the Water Environment Federation.

Higgins, M.J., Beightol, S. Pathak, B. DeClippeleir, H., Al-Omari, A., & Murthy, S. N. (2019). Impacts of feed dilution and lower srt on performance of thermal hydrolysis/anaerobic digestion. *Water Environment Research*, *91*, 386–398.

Higgins, M. J., Murthy, S. N., & Chen, Y. C. (2006). *Understanding factors affecting conditioning and dewatering*. Water Environment Research Foundation Report No. 01-CTS-1.

Higgins, M. J., & Rajagopalan, G. (2017). *Understanding impacts of co-digestion: Digester chemistry, gas production, dewaterability, solids production, cake quality and economics*. Water Environment & Reuse Foundation.

Higgins, M. J., Rajagopalan, G., Miller, A., Brown, J., & Beightol, S. (2017). Effect of food waste co-digestion on digestion, dewatering and cake quality. *Water Environment Research*, *89*, 32–42.

Higgins, M. J., Yarosz, D. P., Chen, Y. C., Murthy, S. N., Maas, N., Cooney, J., Glindemann, D., & Novak, J. T. (2006). Cycling of volatile organic sulfur compounds in anaerobically digested biosolids and its implications for odors. *Water Environment Research*, *78*, 243–252.

Kang, S. J., Olmstead, K., Takacs, K., & Collins, J. (2008). Municipal nutrient removal technologies reference document. EPA Report: 832-R-08-006.

Kim, M., Chowdhury, M. M. I., Nakhla, G., & Keleman, M. (2017). Synergism of co-digestion of food wastes with municipal wastewater treatment biosolids. *Waste Management, 61*, 473–483.

Murthy, S.N. & J.T. Novak. 1999. Factors affecting floc properties during aerobic digestion: Implications for dewatering. *Water Environment Research*, *71*, 197–202.

Murthy, S.N., Novak, J. T., & Holbrook, R. D. (2000). Optimizing dewatering of biosolids from autothermal thermophilic aerobic digesters (ATAD) using inorganic conditioners. *Water Environment Research*, *72*, 714–721.

Navaratnam, N., Topczewski, P., Royer, S., & Zitomer, D. (2011). Blending anaerobic co-digestates: Synergism and economics. *Water Science & Technology*, *63* (12), 2916–2922.

Neave, S. L., & Buswell, A. M. (1930). The aerobic oxidation of fatty acids. *Journal of the American Chemical Society*, *52*, 3308–3314.

Novak, J.T., Park, C., & Abu-Orf, M. (2004). Conditioning and Dewatering of Digested Waste Activated Sludge. *Journal of Residuals Science & Technology, 1*, 45–51.

Novak, J. T., Adams, G., Card, T., Chen, Y. C., Erdal, Z., Forbes Jr., R. H., Glindemann, D., Hargreaves, J. R., Hentz, L., Higgins, M. J., McEwen, D., Murthy, S. N., & Witherspoon, J. (2006). Generation patterns of sulfur containing gases from anaerobically digested biosolids. *Water Environment Research*, *78*, 821–827.

Rajagopalan, G., Higgins, M., Miller, A., Brown, J., Goepfert, R., & Mohammed, M. (2014). *Effect of food waste co-digestion on dewatering, cake quality and digestion parameters*. Proceedings of the Water Environment Federation, 1–18.

Rajagopalan, G., Subramanian, B., & Higgins, M. J. (2022). Quality of biogas derived from wastewater solids and co-digested organic wastes—A characterization study. Project 4892. The Water Research Foundation.

Rittmann, B. E., & McCarty, P. L. (2001). *Environmental biotechnology: Principles and applications*. McGraw Hill.

Sotemann, S. W., Tistow, N. E., Wentzel M. C., & Ekama, G. A. (2005). A steady state model for anaerobic digestion of sewage sludges. *Water South Africa*, *4*, 511–527.

U.S. Environmental Protection Agency. (2021a). *Emerging issues in food waste management: Plastic contamination*. Report No. EPA/600/R-21/116.

U.S. Environmental Protection Agency. (2021b). *Emerging issues in food waste management: Persistent chemical contaminants*. Report No. EPA/600/R-21/115.

Zitomer, D. H., Adhikari, P., Heisel, C., & Dineen, D. (2008). Municipal anaerobic digesters for codigestion, energy recovery, and greenhouse gas reductions. *Water Environment Research*, *80*, 229–237.

5.0 SUGGESTED READINGS

Ekama, G. A., Sotemann, S. W., & Wentzel, M. C. (2007). Biodegradability of activated sludge organics under anaerobic conditions. *Water Research* *41*, 244–252.

Silvestre, G., Illa, J., Fernández, B., Bonmatí, A. (2014). Thermophilic anaerobic co-digestion of sewage sludge with grease waste: Effect of long chain fatty acids in the methane yield and its dewatering properties. *Applied Energy*, *117*, 87–94.

Yenigün, O., & Demirel, B. (2013). Ammonia inhibition in anaerobic digestion: A review. *Process Biochemistry*, *48*, 901–911.

York, R. V., & Magner, J. A. (2009). *Increasing CHP productivity while reducing biosolids volume and climate changing gasses*. Proceedings of the Water Environment Federation, 6639–6651.

Chapter 5

Feedstock Receiving, Processing, and Handling

Zhongtian Li, PhD, PE; Joerg Blischke, PE; William Chen, PE, BCEE; Jonathan Koch; Michael K. Stenstrom, PhD, PE, BCEE; & Christian Tasser, PE

1.0 INTRODUCTION

Organic waste streams, including food scraps, are valuable energy and nutrient resources that have mostly been discarded in North American landfills. Recently, there has been a growing effort from both private and public sectors to collect, process, and utilize this waste/resource as an organic feedstock for co-digestion at water resource recovery facilities (WRRFs) in certain regions of the United States (U. S. Environmental Protection Agency [U.S. EPA], 2022). By converting municipal organics stream into methane-rich digester gas—or biogas—through anaerobic digestion processes, these facilities can harness the renewable energy present in the waste (Zhang et al., 2023).

Diverting organic waste streams from landfilling for beneficial use through co-digestion is a promising pathway, a process that involves the combined treatment of food scraps with other organic waste streams, such as agricultural residues or wastewater sludge (Cheng et al., 2020). Co-digestion not only enhances the efficiency of waste decomposition but also optimizes the production of biogas—a renewable source of energy. However, the success of co-digestion largely depends on effective pretreatment/preprocessing techniques that transform these typically inhomogeneous organic streams for optimal digestion (Abu-Orf et al., 2023).

Preprocessing is paramount to overcome the unique challenges posed by source separated organics (SSO) and organic fraction of municipal solid waste (OFMSW), including inconsistent composition, high physical contaminant levels, and logistical hurdles. In this study, SSOs are defined as biodegradable materials, including food scraps and yard waste, which are segregated from other waste streams at their origin and are processed separately through composting, anaerobic digestion, or other forms of conversion into renewable energy and/or bioproducts. The term *feedstock* is used in Chapter 5 encompassing SSO, OFMSW, and other types of organic wastes that are suitable for co-digestion.

One of the primary challenges in managing feedstock is their inconsistent compositions. Food scraps typically comprise a diverse mix of organic materials, including fruit and vegetable peelings, meat scraps, dairy products, and processed food items. This variability makes it difficult to standardize treatment processes and predict the outcomes of waste management practices. Moreover, seasonal variations, source/location of generation, and cultural differences in dietary habits can further complicate the composition of food waste (Usmani et al., 2021).

Another significant issue with feedstock is the commonly observed high level of physical contaminants present (Szmidt, 2022). These physical contaminants and debris can include non-organic materials such as plastic packaging, produce stickers, gloves, metal cans, cutlery, batteries, glass, and other non-biodegradable/digestible items (e.g., textiles, eggshells, rocks, grit, plastics, and microplastics) that are inadvertently mixed with food waste. The presence of such physical contaminants poses several challenges for waste processing facilities, including operational inefficiencies, premature wear and damage of equipment including conveying systems, and increased processing costs. In addition to physical contaminants, food waste can also harbor chemical contaminants such as pesticides, heavy metals, and other hazardous substances that can impede treatment processes and pose risks to human health and the environment.

Feedstock that contains or have been contaminated by meat and any other animal material may bring concerns of biological contaminations of pathogenic bacteria and virus (Iacovidou et al., 2012). For instance, the wastewater industry must have an animal byproducts (ABP) permit to process food waste. The animal byproducts regulations (ABPR) require the removal of physical contaminants and pasteurization of food waste before co-digestion. Although pasteurization of food waste is not required for co-digestion in the United States, pathogen control of co-digestion feedstock is required to produce Class A biosolids based on the Standards for the Use or Disposal of Sewage Sludge (40 CFR Part 503).

The management of feedstock is further complicated by logistical issues. Food waste is typically generated in numerous locations, including households, restaurants, cafeterias, food distribution centers, supermarkets, and food processing facilities, which necessitates an extensive and efficient collection system (Mabe et al., 2022). The perishable nature of food waste also requires timely and frequent collection to prevent odor issues, pest infestations, and the rapid degradation of organic matter. Additionally as putrescible co-digestion feedstock has a high water content, the transportation to processing facilities must be managed efficiently to minimize greenhouse gas emissions and reduce the overall carbon footprint associated with waste management (Babbitt et al., 2022).

The effective management of feedstock necessitates a comprehensive understanding of its inconsistent composition, high physical contaminant levels, and logistical challenges. Addressing these complexities is essential for developing sustainable and efficient strategies for feedstock preprocessing and co-digestion. Chapter 5 delves into the critical role of preprocessing of co-digestion feedstocks. We will begin by understanding the nature of food waste, exploring its variable composition, physical contaminants that complicate its management. We will explore various preprocessing techniques that can be used to preprocess feedstocks for effective co-digestion.

2.0 FEEDSTOCK TYPES AND QUALITIES

A 2021 report from the U.S. EPA revealed that approximately 1.3 million tons of food waste were co-digested at WRRFs. It's worth noting that the majority of the digested food waste is beverage processing waste digested by dedicated stand-alone digesters, as shown in Table 5.1 (U.S. EPA, 2021). The data suggested a potentially abundant beverage-related feedstock is available for co-digestion when favorable techno-economic factors exist. The survey data also showed that various types of feedstocks were being accepted, including leftovers from restaurants and food services, municipal or residential scraps, surplus produce from grocery stores, and excess retail items. The quality of the input materials is crucial for the success of co-digestion projects, as it impacts both capital investments and operational costs. It is recommended to reference Chapter 3, "Feedstocks Evaluation and Selection," for additional information.

The diverse composition and quality characteristics of feedstock from different regions, along with potential changes over time due to source variability, pose a major challenge for those involved in handling, processing, and regulating this material. As illustrated in Table 5.2, a wide range of sources contributing to feedstock complicates the sampling and characterization process, making it difficult to obtain reliable and consistent results. Furthermore, co-digestion is becoming more closely linked to environmental concerns, including initiatives like California's carbon intensity score that promote low-carbon fuel production through biogas utilization.

TABLE 5.1 Total Amount of Food Waste Processed in 2020 and 2021 by U.S. EPA Survey Respondents

Digester Type	Reported Amount Processed in 2020 (tons)	Reported Amount Processed in 2020 Without Beverage Processing Waste (tons)	Reported Amount Processed in 2021 (tons)	Reported Amount Processed in 2021 Without Beverage Processing Waste (tons)
Stand-alone digesters	14 658 400	55 700	15 530 100	189 000
On-farm co-digesters	8600	8400	33 700	31 700
Co-digestion systems at WRRFs	1 139 800	1 129 100	1 323 600	1 286 300
Total	15 806 800	1 193 300	16 887 400	1 507 100

TABLE 5.2 Co-digestion Feedstock Types

Type of Feedstock	Sources
Preprocessed Bioslurry	Commercially produced feedstock slurries manufactured by waste management practices.
Residential Source, Post-consumer	Food waste collected from dedicated organic waste bins by municipal waste collection services from single and multi-family units.
Commercial Source, Post-Consumer	Food waste collected from commercial and institutional services such as hospitals, correction facilities, universities, central kitchens/cafeterias, restaurants.
Commercial Source, Pre-consumer	Food manufacturers, food packaging facility and warehouse/food distribution centers, grocery stores, food preparation services.

Food waste is highly biodegradable and typically exhibits a higher rate of volatile solids reduction (80%–90%) compared to traditional municipal wastewater sludges (50%–65%). Consequently, when a co-substrate is introduced into anaerobic digesters, it leads to an increase in biogas yields with only minimal impact on biosolids production. The elevated proportion of volatile solids in food waste also renders it more readily degradable than solid components found in conventional wastewater sludge. In general, when blending co-feedstocks, the possible outcomes can be synergistic, antagonistic, or neutral based on methane production that is greater than, less than, or the same as that observed when each material is digested individually (Zitomer et al., 2008). Certain utilities have observed that co-digestion stimulates biological activity and enhances the overall degradation of volatile solids and reduces biosolids output as opposed to relying solely on wastewater sludges for digestion purposes (Nghiem et al., 2017).

The properties of feedstocks can significantly influence various aspects including solids treatment, operational processes, equipment functionality, product quality assurance, and compliance with regulations. Standardizing criteria and implementing testing protocols for evaluating feedstocks is essential to ensure regulatory compliance and the efficiency of end-products (Rauch-Williams et al., 2023). By providing clarity and guidance to stakeholders involved in the supply chain, these measures contribute to the overall success and sustainability of organic waste management schemes in the sector. This study offers an overview of the essential characteristics of feedstock that are important for safeguarding mechanical equipment including conveying systems and infrastructure, managing odors, optimizing anaerobic digestion efficiency, ensuring biosolids quality, maximizing energy recovery, enhancing biogas quality, and maintaining effluent quality.

A comparison in Table 5.3 highlights the general qualities of different SSO feedstock categories based on factors like biogas generation potential and physical contamination levels (Rauch-Williams et al., 2023). It is crucial to recognize that variations in quality may exist within specific SSO

TABLE 5.3 Co-digestion Feedstock Qualities

Feedstock Characteristics	Commercial Source, Pre-consumer	Commercial Source, Post-consumer	Residential Source, Post-consumer	Preprocessed Bioslurry
Biochemical Methane Potential (BMP)	BMP is highly dependent on the type of SSO. High BMP if with a significant portion of FOG.	BMP is comparable to preprocessed residential post-consumer SSO.	BMP may be lowered by mixing with light green waste.	BMP can be customized based on the proportion of the blending feedstocks.
Physical Contamination Level	The level of contamination is determined by the commercial processes generating SSO. The presence of plastics, utensils, glass, bones, wrapping films, produce stickers, and packaging materials are typical without preliminary treatment.	The level of contamination is usually lower than residential post-consumer SSO. The presence of plastics, utensils, glass, bones, wrapping films, and packaging materials are typical without preliminary treatment.	The presence of plastics, utensils, paper, cardboard, glass, metals, stones, ceramics, textiles/clothing, and other non-organic/non-digestible waste are very common.	The level of contamination is controlled to meet site specific specification.
Overall Quality	Quality depends on the source of the SSO.	Level of contamination is usually higher than commercial pre-consumer SSO. High variation of the SSO quality due to diverse sources.	The level of contamination is usually higher than commercial post-consumer SSO. High variation of the SSO quality due to diverse sources.	The level of contamination is usually low because of pretreatment. Usually well mixed and blended.

feedstock types, necessitating consideration of this variability when developing sampling strategies for analyzing hauler deliveries. However, these differences may not necessarily impact on the suitability of material for anaerobic digestion. It is also important to note that the presence of contaminants may not hinder the digestion process.

3.0 UNIT PROCESSES FOR FEEDSTOCK PRETREATMENT

Typically, SSO needs to undergo pretreatment/conditioning before it is suitable for anaerobic digestion or co-digestion. There are three types of pretreatment processes that can be applied: physical/mechanical, thermal, and biological. All three types will be described in the following sections. As physical/mechanical pretreatment is most applied, this process is described in more detail.

3.1 Mechanical Pretreatment

3.1.1 Unit Processes for Feedstock Pretreatment

This process commonly involves removing impurities, reducing particle size, adding process water, and blending different materials to improve chemical properties by adding additional chemicals or nutrients. Current research has focused primarily on processing SSO through methods like composting or anaerobic digestion, rather than evaluating the overall impact of waste composition on reducing food wastage. While some studies have examined variations in characterizing food waste, there has been limited exploration of regional disparities in food waste generation, which could significantly affect how effectively these materials are managed and treated. As a result, industry has been relying on broad assumptions to address physical contamination issues.

Recoverable rejects from SSO processing can include hard and film plastics, cardboard, paperboard, ferrous, and non-ferrous metals. The level of physical contamination in wasted materials can vary widely, from minimal instances like spoiled grains to highly contaminated cases where edible discards collected at the source may contain non-processable substances. Municipal solid waste can contain high levels of indigestible elements such as soils, grit, and inactive domestic refuse (Abu-Orf et al., 2023). Even factory-based food residues may be contaminated, with items like spoiled breakfast cereals packaged in plastic receptacles. Canned goods can also pose contamination risks, as their cans may be made of tin or steel, labels can easily peel off, and the contents may separate from any enclosed liquids (Carol et al., 2019).

Common contaminants found in discarded food items include plastics used for packaging, such as polystyrene, films, and metals like cans and cutlery. Fine articles like teabags, stickers, paper towels and (often non-biodegradable) wipes, multi-material bread bags, coffee filters and pods, and diapers can also be present in food scraps, posing challenges and possible environmental risks. It is crucial to address these contamination issues in the preprocessing of SSO.

Hence, digesting or co-digesting SSO with a high variability in composition in a low-solids digestion process requires physical/mechanical pretreatment steps for contaminant removal and homogenization. While the used pretreatment process is mostly site specific tailored to the specific quantity and quality of the processed organic feedstock(s), the following simplified process schematic (Figure 5.1), represents a general arrangement of the pretreatment train. It is typically comprised of three sequenced Process Unit (PU) steps (with an off-site, on-site, or combined off-site/on-site setup): an Initial PU, followed by a Core PU and concluded with a Polishing PU step.

Depending on the feedstock characteristics and deployed pretreatment technologies, the pretreatment system may be comprised of

- Initial PUs: bag opening, shredding, crushing, screening/separation (e.g., separation of bulkier or non-digestible items; removal of ferrous and non-ferrous metals);
- Core PUs: depacking (e.g., hammermill), pulper, waste separation press; and
- Polishing PU: settling tank, dynamic cyclone, hydrocyclone, screen.

The actual configuration is, as stated, project specific and may include multiple or just one or two PUs.

3.1.2 Feedstock Receiving and Storage

Although procedures such as intermediate storage, including buffer storage and bunkering, are not typically considered pretreatment steps, they

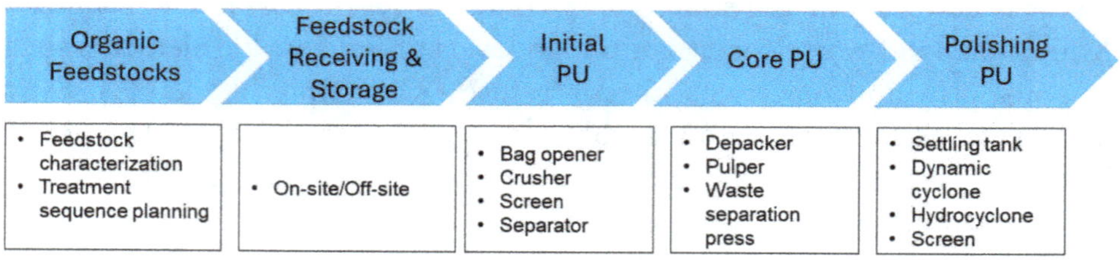

FIGURE 5.1 General Mechanical Source Separated Organics Pretreatment Train Arrangement

play a crucial role in the process by influencing the physical properties of organic materials such as perishable food waste along with odor management considerations. The reception of bulk waste can lead to layering and compaction, increasing bulk density and impacting factors like viscosity.

In large-scale operations, industrial facilities must carefully plan their reception areas configurations to align with SSO collection and delivery schedule (typically 5 to 6 days a week), pretreatment and processing schedule (typically 5 to 7 days a week), and associated storage and processing capacities and capabilities. This ensures efficient treatment of SSO, minimizing the risk of energy losses due to aerobic biodegradation or putrefaction, along with facility permitting requirements and environmental compliance. Delays, especially when utilizing crane-operated deep bunkers, can result in decreased energy value at reception points during the pretreatment phases.

Another aspect to be considered when setting up the SSO reception, storage, and pretreatment system is the selection of location(s). There are three pathways when it comes to location:

1. **Off-site:** The collected organic material is received and pretreated off-site. The produced organic slurry—or paste—is shipped via tanker truck or trailer truck to the WRRF. The received slurry—or paste—may then undergo further conditioning/polishing along with temporary storage before digestion or co-digestion.
2. **On-site:** The collected organic material is received and pretreated on-site at the WRRF, and the produced slurry is temporarily stored before digestion or co-digestion.
3. **Combination off-site and on-site:** Depending on the type of feedstocks/their characteristics, it is also possible to have a combination of these two options. For example, fats, oil, and grease (FOG) and other liquid high-strength wastes may be received and conditioned directly on-site at the WRRF; whereas collected solid/non-pumpable SSO may be pretreated off-site (e.g., at a waste transfer station or material recovery facility) before arriving as an organic slurry—or paste—where it may undergo further polishing.

Which pathway is selected is project specific. For example, the WRRF is not able to receive SSO due to site constraints or operational concerns but is well positioned to receive a clean, pumpable bioslurry that is produced off-site by a third-party waste management firm.

As a pumpable bioslurry has a typical total solids (TS) content ranging between 8% and 17% (thus containing significant amount of water), it is advisable to keep the hauling distance within a 50-mile radius or less. The following sections provide a high-level overview of commonly used PUs under each PU category.

3.1.3 Initial Process Unit Step

The initial process units (initial PUs) are units that are positioned at the beginning of the process train. The processes conducted by these PUs prepare the incoming SSO for the subsequent treatment in the core PUs, the next step in the process train. The PUs in the initial PU step may include initial size reduction, separation or removal of certain content including non-digestible debris, and cutting bags open to expose their organics content. The process units to be discussed in this section, that is, bag opener, shredder/crusher, screening, and metal separation are treatment steps that may or may not be deployed before the subsequent core PUs. It is dependent on the type and composition of the SSO as well as the specific configuration of the subsequent core PUs.

3.1.3.1 Bag Opener

Commonly, SSO is delivered in bags or contained within packaging material. Liberating the organics from any form of containment improves their recovery in the subsequent separation steps. However, the size reduction of the feedstock, for example via crusher, may increase the effort required to separate out different materials. Waste bags (fossil fuel based or biodegradable), film wrapping, and pouches are fed directly into an integrated chute to be opened and loosened up before transferring the material for subsequent stage of treatment. The bag opener is designed to tear open all forms of packaging and containers allowing the contents to be released without shredding/size-reducing the material.

3.1.3.2 Mechanical Shredding and Crushing

Reducing the particle size of anaerobic digestion feedstock can enhance the surface area of waste, making it more accessible to microbes and promoting biogas production. This process may also improve the manageability and consistency of waste. Particle size reduction is a widely used pretreatment method for SSO destined for anaerobic digestion or co-digestion, known to improve substrate utilization and COD solubilization (Bong et al., 2018). However, excessive reduction may lead to anaerobic digestor (AD) instability, causing premature acidification due to small particle sizes. While mechanical pretreatment may not significantly increase biogas production from easily degradable food waste, it is beneficial for materials rich in cellulose and hemicellulose such as straw or vegetable waste. Crushers and shredders come in different configurations. Examples are

- A dual-shaft shredder that is comprised of a large shredding compartment, two long, horizontal counter-rotating shredding drums, and a unique tooth shape that combines to process bulkier material such as large stumps, whole trees, and SSO.

- A single vertical rotating shaft equipped with one or more attached rotating chains, hammer(s), or teeth encapsulated in a cylindrical drum to break down the cell structure of the input material where the input material is defibrated rather than chopped up thus producing a more homogeneous substrate with a large specific surface area.

There are various knife shredders and crushers available for particle size reduction, but caution is advised as reducing particle size can introduce contaminants like plastic fragments, compromising the quality of input materials and the integrity of post-digestion digestate. The presence of inert materials in the digester can pose challenges, such as reduced capacity efficiency and potential operational risks (Abu-Orf et al., 2023).

3.1.3.3 Screening and Separation

There are typically two types of screens used as initial PUs in an SSO pretreatment train: star—or disc—screens and drum/trommel screens. The process of star or disc screening involves the use of one or more screen decks that move the material to be separated (Figure 5.2). Rotating shafts with closely spaced "stars" or discs positioned perpendicular to the material flow direction convey materials away from the input point. Coarse-grain materials are sorted out when they are unable to pass through the gaps between the stars or discs. Typically, the material that does pass through falls onto a conveyor and then onto a fine screen deck equipped with smaller stars or discs for further separation into fine and medium fractions. The configuration of a standard two-stage star or disc screen allows for the adjustment of particle size separation fractions by changing the rotary speed of the star or disc shafts. More advanced setups provide control over grain size using frequency converters based on star geometry.

Drum screens, also known as trommel screens, are commonly used in mechanical sorting in waste and recycling industries (Figure 5.3). They are versatile and can be used for screening sand, gravel, soil, as well as separating inert and semisolid organic fractions. A drum/trommel screen consists of an inclined cylinder mounted on rollers with holes or slots of specific sizes punched or cut along its side, determining what can pass through for collection below, usually onto a conveyor belt for further transport.

A more recent addition to screening PU category is the spiral shaft screen/separator marketed under the trade name SPLITTER®. The screening and

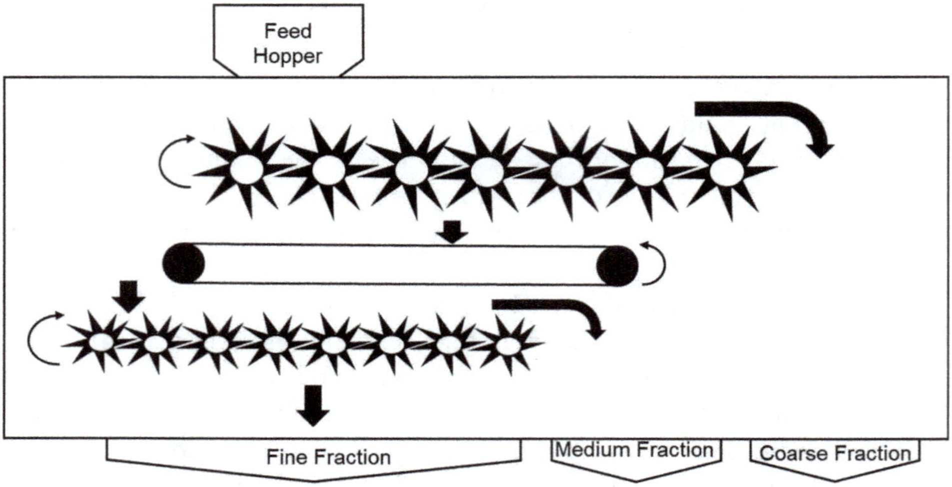

FIGURE 5.2 Schematics of Star Screen

separation process is based on the material's physical properties: size, weight, and shape where the input stream is separated into up to three fractions.

The single side driven and mounted spiral separator combines a roller track effect with the principle of an Archimedean screw: the spiral shaft guides the material laterally, while the roller track effect at the same time transports the material in conveying direction. Compared with a conventional screen the spiral has an open end where the gap "runs out" for one of the fractions to be separated out and discharged. Three different screen

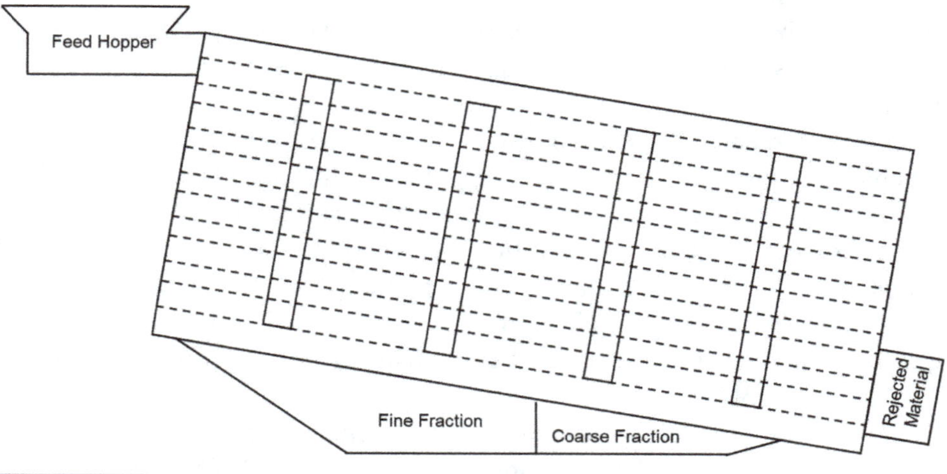

FIGURE 5.3 Schematics of Drum Screen

deck configurations are available to tailored to the feedstock characteristics, throughput, and desired fractions to be separated.

Ferrous and non-ferrous removal, often positioned as part of the initial PUs, is a process before the core PUs to better separate the non-digestible or non-organic fraction (i.e., metals) from the digestible. Magnets used in separation recover ferrous (iron) and eddy current separation is used for separation of non-ferrous metals such as aluminum. The separated metals may undergo further treatment for material recovery on-site or off-site. The organic-rich stream still contains contaminants necessitating further mechanical treatment downstream.

3.1.3.4 Manual Sorting

Sorting and separation technologies are commonly used to eliminate large, visible contaminants from the waste stream before advancing through the treatment process. The presence of oversized waste components poses a risk of equipment blockages and hinders downstream treatment efficiency. Manual sorting operations like picking lines may still play a crucial role in extracting recyclable (positive sorting) or undesired materials (negative sorting) from the waste flow, serving as an initial pretreatment method adopted by numerous full-scale treatment facilities (Smith, 2005) Additionally, separation technologies have been integrated into full-scale facilities to extract recyclable or inorganic substances from the waste stream, thereby concentrating organic matter within processed SSO (Pognani et al., 2010).

3.1.4 Core Process Units

The core PUs are process units that are considered the centerpieces of the process train that achieve the following three objectives:

- (Further) removal of physical contaminants,
- Further particle size reduction, and
- Homogenization/conditioning (mixing) to a greater or lesser extent.

It is in the core PU that the generation of the bioslurry takes its initial form. It is in this step that the target TS content of the produced bioslurry is commonly achieved, and the TS content ready to be fed to the subsequent Polishing PU or digester typically ranges between 5% and 15%.

The separation of physical contaminants and homogenization (the latter to a greater or lesser extent) is relying on physical/mechanical techniques aided in some cases by the addition of process water. The physical/

mechanical techniques may rely on rotating equipment (paddles or hammers on one or more rotating horizontal or vertical shaft or augers/agitators) or pressing or squeezing equipment using a piston or screw-type motion within a chamber with wall perforation. Depending on the type of core PU used, it can be either a continuous, semi-continuous, or batch/discontinuous process. Depending on the core PU, process water is added to facilitate the separation of physical contamination, homogenization of the organic feedstock, and removal of residual organics from physical contaminates before discharge. The core PUs used in the industry may be one of the technologies listed in Table 5.4 below. The core PUs will be discussed in detail in dedicated sections of this chapter.

3.1.4.1 Depacker/Separation Mill

A variety of size-reduction and separation equipment falling under the depacker category that is using a hammer or paddle mill principle available to treat SSO. Due to the hammer or paddle mill principle, this type of equipment is also known as hammer mills. These depacker/separation mills play a dual role in reducing particle sizes and separating digestible components from non-processable substances in incoming SSO streams (Figure 5.4). This equipment is designed with swinging hammers or fixed paddles mounted on rotating shafts inside drums to impact material and achieve desired separation results based on specific settings for different types of input materials (Coker, 2021).

Source separated organics entering the depacker is hit by sufficient force that crushes or tears the material to be processed with a velocity so that it does not adhere to the rotating hammers or paddles. The size of the waste is further reduced by being struck between the hammers and encapsuling screen with specifically designed holes or slots surrounding the hammers. This process of cutting and striking continues until the required size of the material is achieved and escapes through the screen holes or slots. To generate a pumpable waste slurry and a "cleaner" reject stream, process water is typically added in many of the depacker applications.

TABLE 5.4 Major Core Process Unit Technologies

Technology	Typical Mode of Operation	Mode of Action
Depacker	Continuous	Breaking, crushing
Pulper	Batch	Shearing, tearing
Waste Separation Press	Semi-continuous to continuous	Pressing, squeezing

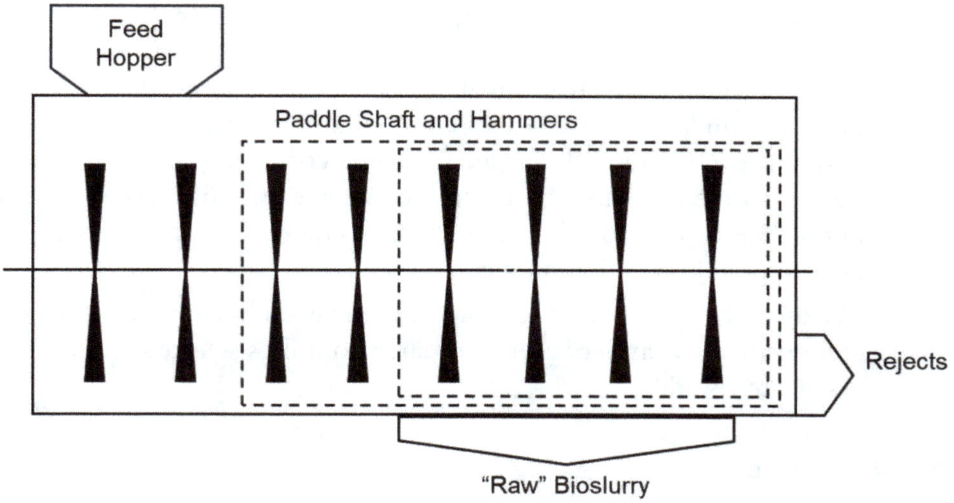

FIGURE 5.4 Schematics of Mechanical Shredding and Mill Separation

The produced bioslurry may then be transferred to polishing PUs for further processing. The rejected steam may be directed to a dewatering step that may also include a rinsing/cleaning step to remove organics attached to the physical contaminants.

The following provides a brief description of some of the key advantages and challenges/disadvantages for depacker process units receiving and treating a wide variety of SSO.

Advantages:

- Can process SSO with poor "holding structure," for example, SSO with a sloppy consistency and high-water content.
- Continuous process.
- Can process SSO without the need to add process water.
- Residuals organics in rejected streams are typically low (especially when process water is added).
- Relatively low in capital cost.
- Several equipment suppliers to choose from (>10).

Disadvantages:

- Not well suited to processing feedstocks that contain glass (e.g., bottles, containers) and stones.

- Depackers tend to crush, smash, shatter, or tear physical contamination into smaller to very small pieces like glass and plastic fragments. Caution is advised as these small pieces compromise the quality of the bioslurry, and, if not removed before entering the digestion process, may adversely affect the bioslurry piping/conveying system and the quality of the digestate. Furthermore, the presence of inert materials in the digester can pose challenges, such as reduced capacity efficiency and potential operational risks.
- Residual organics in the reject can be high depending on the feedstock composition, what depacker or separation mill is selected and if process water is added.

3.1.4.2 Pulper

Developed in the mid-1980s in Europe, solid waste pretreatment with pulper technology is based on designs from the pulp and paper, and ceramic industry. The pulper design tailored to the specific feedstock characteristics has seen several evolutions and configurations over the past three decades. Pulper technology is a discontinuous/batch process and is often offered as part of a complete pretreatment system (core PU followed by polishing PU).

In the pulper, SSO and process water are added and mixed. The high-speed rotating auger, mounted on a rotor at the bottom or top of the cylindrical vessel, transforms, thanks to the auger's created sheer forces, pulpable and friable materials into a bioslurry. There are various pulper configurations with associated equipment to separate and remove undissolved debris. Figure 5.5 below illustrates a pulper design example by BTA International GmbH, distributed by CCI Bioenergy in North America.

This undesired debris, such as heavy and light fractions, may be removed separately or combined. First, separate removed fractions: heavy fractions, such as metals, tins, cans, glass, and other non-pulpable and non-friable materials are rejected from the side of the pulper vessel. The heavy rejects pass down from the pulper vessel to a collection system that gathers the heavy rejects. The light fractions, such as plastics, textiles, and woody biomass, rise to the surface and float on top of the mixture inside the pulper tank. These floating materials would become light rejects that are removed and collected. Both the heavy fraction and light fraction rejects are then collected separately as rejects and sent to the subsequent rejects management system. Second, the entire pulped SSO with its debris is directed from the pulper vessel to an attached separator unit that then separates the bioslurry from the undesired debris sent to the subsequent rejects management system. Although other pulper design configurations exist, they share a common process design objective that is transforming the inhomogeneous and contaminated SSO into a homogeneous bioslurry with low residual physical contamination.

FIGURE 5.5 Example of a Pulper Showing the Auger and Drive Mounted at the Bottom of the Pulper Vessel (Reprinted with permission from CCI Bioenergy. CCI BioEnergy, Inc., is the exclusive partner for BTA International GmbH for Canada and the United States.)

The produced bioslurry may then be transferred to a polishing PU for further processing. The rejected steam is commonly directed to one or more dewatering steps that may also include a rinsing/cleaning step to remove organics attached to the physical contaminants. The following provides a brief description of some of the key advantages and challenges/disadvantages for pulper PUs receiving and treating a wide variety of SSO.

Advantages:

- Can process SSO with poor "holding structure," for example, SSO with a soupy consistency and higher water content.
- Can process glass containers (up to a certain amount) and feedstocks that contain stones.
- The pulper agitator induces sheer forces to disperse/pulp the biodegradable material, reducing the chance of tearing or smashing the non-biodegradable materials into smaller pieces.

Challenges/disadvantages:

- Requires the addition of process water.
- Discontinuous/batch process.
- Relatively high in capital cost.

- Residual organics in the reject may be elevated depending on the feedstock composition, what pulper design is used and how it is operated.
- Limited number of system suppliers to choose from (<5).

3.1.4.3 Waste Separation Press

When it comes to mechanical compression of SSO for separation purposes, applying active pressure against a screen or perforated matrix can effectively divide SSO into mostly fine organic particles and rejects. This process differs from star/disc/drum/spiral shaft screenings as it does not involve medium to fast rotating action but rather slowly forces the SSO through either rotational compression (continuous process) or direct compaction (semi-continuous process).

The piston or plunger-type waste separation press (WSP) typically uses hydraulic pressure to extract the liquid fraction of the infeed material through a perforated screen or matrix when pressed against a plate or gate that is opened and closed to discharge the residual stream as reject. The screw-type WSP separates liquid from solids by continuously rotating the input material against a circumferent perforated screen through which the organic rich liquid is passed through while the solids are continuously discharged at the end of the screw that works against a pressurized plate or cone with a gap opening.

Separation of SSO through mechanical pressing is a relatively new approach that was first introduced in the early to mid-2000s. The typical design relies on a WSP via piston, plunger, or screw that separates the waste into a wet fraction for biological treatment and a fraction of dryer residuals for further material recovery, composting, thermal treatment, landfilling, or other uses.

Depending on the type of organic feedstock, it is possible to feed the WSP without upstream mechanical contaminant removal, crushing or milling and screening of organic waste before the biological treatment.

The following provides a brief description of some of the key advantages and challenges/disadvantages for WSP PUs receiving and treating a wide variety of SSO:

Advantages:

- Able to process infeed containing glass (e.g., containers; jars) and stones (semi-continuous WSP type).
- Can process SSO without the need to add process water (semi-continuous WSP type).

Challenges/disadvantages:

- Limited to no ability to process SSO with poor "holding structure," for example, SSO with a soupy consistency and higher water content (depending on WSP type).

- Semi-continuous process.
- Residual organics in reject can be high depending on the feedstock characteristics and WSP selected.
- Medium to high in capital cost (depending on the type of WSP).
- Limited number of system suppliers to choose from (<10).

3.1.4.4 The "Dragging Factor"

Depending on the type of core PU used, there is a chance that organics are carried out with the rejected fraction to a lesser or greater extent. This phenomenon is also described as a "dragging factor." As explained by Centemero and Confalonieri in 2024, the dragging factor is caused by the sticky nature of SSO, which tends to stay attached, or "glued," onto the reject material such as plastics. This in turn will reduce the amount of organics available for AD (and loss biogas yield—and potential revenue) and increase the amount of reject stream (increase in disposal cost) during the separation process. According to Centemero and Confalonieri, the observed dragging factor, on average, is 2.75. Thus, 4% of physical contaminants, for example, result in 11% of final rejects when considering the organics "sticking" to the contaminants.

Hence, careful consideration is given to the selection of the core PU and associated equipment to maximize the amount of organics extracted from the SSO stream while minimizing the amount of residual organics in the reject stream.

3.1.5 Polishing Process Units

The bioslurry produced by the core PUs is transferred to the polishing PUs for further polishing with the goal of removing physical contaminants, mostly in the form of debris, which remained after the initial PUs and core PUs. This step, very often overlooked in the initial pretreatment train design, has proven essential to sustain reliable operation of the pretreatment train and subsequent digestion to

- Protect pumps and piping systems from blockage and premature wear,
- Minimize the formation of a floating blanket in the digester, and
- Reduce settling out of debris in the conveying system and digester that compromises its capacity.

The following sections provide brief descriptions of polishing PUs that commonly form an integral part of a complete pretreatment train. Often, polishing PUs are added retroactively downstream of the core PU after it was found that the installed pretreatment train did not provide adequate removal of residual debris carried with the bioslurry, causing operational upsets such as premature equipment wear and shorter than anticipated maintenance cycles.

3.1.5.1 Settling Tank

Various circular and longitudinal settling tank designs are on the market to capture and remove fast settling solids/debris from the bioslurry by settling out at the bottom of tank. The flow velocity in the settling tank is reduced to allow solids/debris to settle out. Settled solids/debris are collected at the bottom of the tank to be removed periodically while the desired bioslurry flows to the next stage of treatment.

Practice has shown that using settling tanks that are fed with bioslurry with elevated TS concentrations, for example, greater than 15%, may not be as effective in settable debris removal as the viscosity of the bioslurry hinders debris sedimentation without further dilution. Debris started to settle in the digester due to the decrease in viscosity decrease of the digester content as digestible material converted into digester gas.

3.1.5.2 Dynamic Cyclone

A more recent polishing PU brought to market is a so-called "dynamic" cyclone downstream of the core PU to remove lighter debris (plastic film, paper, and fibers), as shown in Figure 5.6. It is comprised of an interior shaft mounted paddle assembly that is rotating within an encapsulating perforated screen to remove the light debris. The screen size is selectable. This PU is then typically followed by a hydrocyclone for grit removal.

3.1.5.3 Hydrocyclone

A commonly used equipment for fine dense debris removal is a hydrocyclone. It is designed to remove solids, such as grit (sand, metal pieces, glass and ceramic particles, and other heavier organic and non-organic fine debris). The bioslurry enters the top cylindrical portion of the hydrocyclone tangentially and high centrifugal force inside the hydrocyclone thrusts dense particles carried with the bioslurry towards the interior wall of the hydrocyclone. These dense particles slide along the inside wall (cylindrical section followed by conical section) of the hydrocyclone and exit as a concentrated slurry, rich in inorganic solids as the hydrocyclone underflow stream. The hydrocyclone separates the fast-settling solids (both organic

removal.

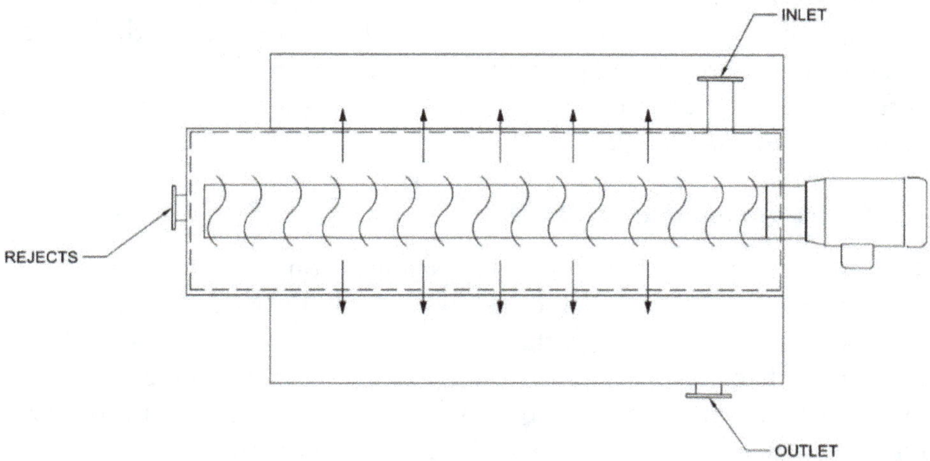

FIGURE 5.6 Simplified Dynamic Cyclone Schematic

and inorganic) from the bioslurry based on density, and the underflow is then typically directed toward a classifier pipe, grit box, and grit washer. The remaining bulk of the now de-gritted bioslurry containing light or dissolved or suspended organic matter and some fine mostly inorganic debris (typically <2mm in size) exit the hydrocyclone through the central riser pipe at the top as part of the overflow bioslurry flow. Practice has shown that the grit removal efficiency can be enhanced by either arranging two or more hydrocyclone in series or to channel the bioslurry two or more times across the same hydrocyclone. Attention must be given to the TS, viscosity, flow rate, and velocity of the bioslurry infeed for the design and selection of the grit removal system as these parameters are affecting the performance of the hydrocyclone. There are several equipment suppliers on the market that offer the hydrocyclone as a stand-alone unit or as part of a complete SSO pretreatment system.

3.1.5.4 Screen

Less common is the use of screens such as disc screens as polishing PUs. They typically target the removal of lighter debris through a screen with desired opening size. The captured lighter debris is retained for further treatment or recycling while the bioslurry that may contain residual physical debris is conveyed to the next stage.

3.1.5.5 Grinder and Macerator

Grinder and macerator achieve size reduction depending on their designs of one or more rotating blades (Figure 5.7). It is worth noting that grinders usually offer more cutting power while machines produce more consistent cuttings. Both types of equipment offer protection of downstream equipment and process piping.

3.1.6 Ultrasound and Sonication

The use of ultrasound, also known as sonication, is a technique used to disrupt cells, commonly used for cell lysis in laboratory environments. This method utilizes pulsed, high-frequency sound waves to agitate and break down cells, bacteria, spores, and other small organic substances. By inserting a vibrating probe into a liquid suspension, homogenized food waste, high-frequency sound waves are produced. The mechanical energy from the probe initiates the creation of tiny vapor bubbles that implode, generating shockwaves throughout the targeted material (Usmani et al., 2021). Ultrasonic treatment is usually applied in short bursts and is the most effective for small-volume batches. Research indicated that sonication notably increased

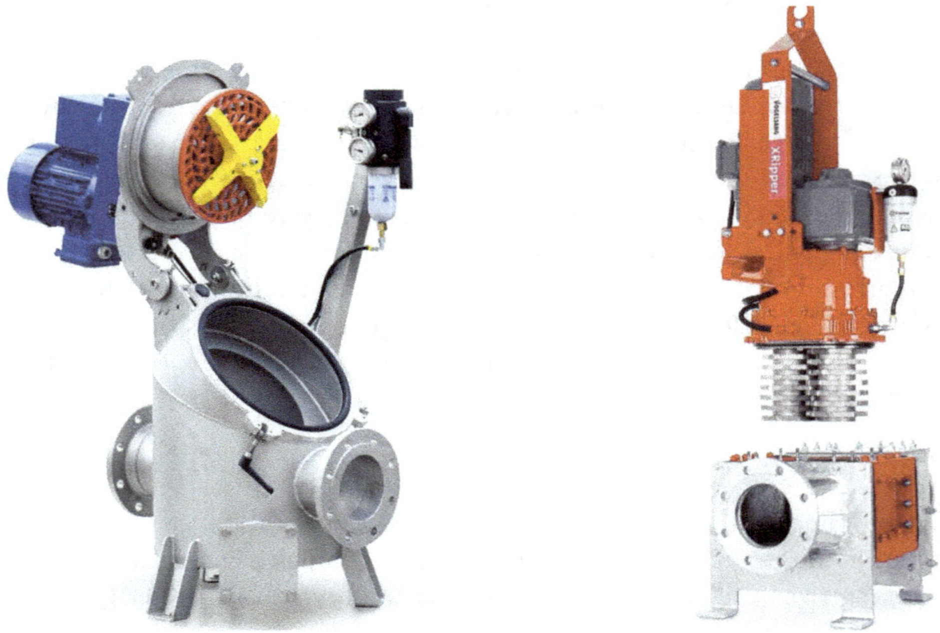

FIGURE 5.7 Macerator (Left) and Twin Shaft Grinder (Right) (Reprinted with permission from Vogelsang)

biogas yield by approximately 27% (in single-stage AD) and energy yield by around 13% (in two-stage AD) compared to untreated controls in laboratory-scale digesters (Elbeshbishy & Nakhla, 2011).

3.2 Thermal Pretreatment

Various organic waste types have been subjected to thermal pretreatment techniques to modify their structure by disrupting intermolecular bonds, facilitating the release of soluble organic monomers/dimers. This process increases the accessibility and biodegradability of these compounds by anaerobic bacteria, thereby improving the kinetics of anaerobic digestion and enhancing reactor efficiency. The efficacy of thermal pretreatment is influenced by factors such as heating temperature and duration, with experimentation showing that a range of temperatures and treatment times can optimize methane yield from food waste (Kovacic et al., 2024).

The main goal of thermal pretreatment is to increase biogas production by enhancing waste particle solubility, leading to energy recovery within the system. However, an identified drawback is that the optimal temperature for improving biogas production may differ from that maximizing waste particle solubility. Integrating thermal pretreatment with screening, depackaging, or separation technologies has been suggested to enhance the overall effectiveness of waste treatment before anaerobic digestion (Szmidt, 2022).

Hydrothermal carbonization is a thermal conversion process that transforms food waste and its packaging materials into a valuable, energy-rich resource. Researchers conducted experiments using food waste collected from local restaurants, subjecting it to different temperatures (225, 250, and 275 °C) and varying solids concentrations to analyze the impact of process conditions on the properties and composition of the carbonization products (Usmani et al., 2021). The study revealed that a significant portion of the initial carbon remains within the solid phase under the conditions tested, with the initial solids concentration playing a key role in carbon distribution due to increased compound solubilization. Interestingly, changes in reaction temperature had minimal effects on carbon distribution. Furthermore, the presence of packaging materials was found to influence the energy content of the resulting solids, with higher proportions of packaging materials leading to decreased energy content due to their lower energetic retention capabilities. While carbonization of food waste and packaging materials also yielded positive energy balances, the energy requirements for post-processing the solids were noted to be significant. The study emphasized the need for a comprehensive life-cycle assessment to fully evaluate the advantages of carbonization beyond just process energetics.

3.3 Biological Pretreatment

Various biological methods of pretreatment are used to convert complex organic matter into simpler compounds through hydrolysis. Enzymatic treatment targets proteins and carbohydrates, yielding amino acids and monosugars, respectively, while fungal treatment releases carbohydrates-active enzymes like cellulase and hemicellulose, leading to the breakdown of complex organic matter (Sagar et al., 2024). The combined action of fungi and enzymes often results in the generation of additional beneficial byproducts, showcasing the potential for innovative approaches in waste management.

While composting reduces the energy content of substrates, it should be limited to preserve energy potential or biomethane yield during subsequent AD stages. Studies indicated that aerobic pretreatments could boost methane production by up to 40% (Jain et al., 2015). Different references discuss strategies to enhance hydrolysis in anaerobic digestion, in which biological pretreatments involving enzymes and bacteria help break down sludge, resulting in increased biogas production (Ariunbaatar et al., 2014). However, longer retention times may be needed for effectiveness, potentially causing more odors. Another study explored garbage disposers and biological solubilization using circulation tanks and reactors for food waste treatments, emphasizing the importance of water levels and oxygen availability under aerobic conditions facilitated by an aerated tank system with continuous water supply. Challenges related to microbial competition within food waste can impact the efficacy of biological agents used for treatment processes, including extended retention periods and high costs associated with enzyme use. Limited literature is dedicated to using biological pretreatments specifically for food wastes, suggesting the need for further research before widespread adoption in treatment facilities becomes feasible.

Drylet specializes in manufacturing products aimed at improving biogas production efficiency and reducing solids at WRRF, such as the Bio React AD (Abu-Orf et al., 2023). This product is designed as a biocatalyst supplement to enhance microorganism populations in digesters, promoting improved hydrolysis and increased biogas yields. Notable benefits of this product include compatibility with no specialized new equipment requirements, although empirical evidence of its application toward SSO-containing feedstock treatments remains limited, despite Drylet indicating the potential utility of BioReact, particularly for food waste management strategies.

3.4 Nutrient Balance

Enhancing the efficiency of anaerobic digestion using pretreatment technology offers a range of possibilities for integrating supplementary substances into the process flow. These additions may include essential micronutrients, enzymes, and chemicals like ferric chloride, which is commonly used for biogas desulfurization according to studies (Banks et al., 2012). Implementation methods vary from techniques such as dosing hoppers connected to conveyors to more direct approaches, like injecting these substances into treatment vessels such as thermal heat tanks. Research emphasizes that enhancing the bioavailability of micronutrients can be achieved through heat pretreatment, highlighting the importance of incorporating them early in the process flow (Almansa et al., 2015).

Maintaining a balance of micro- and macro-nutrients in the feedstock is crucial, especially when there are disparities in nutrient levels within the digesters. Different types of microorganisms have varying requirements for growth and organic breakdown, making it essential to ensure an adequate nutrient supply. A study points out that insufficient presence of essential micronutrients like cobalt, iron, nickel, and sulfide in the feedstock can lead to inefficiencies, hampering the waste degradation capacity of the system (Bardi et al., 2023). To address these limitations, it is vital to provide a balanced mix of both macro- and micro-nutrients necessary for efficient anaerobic digestion. Regular evaluation of feedstock samples is recommended to assess their nutritional adequacy, enabling proactive measures to prevent disruptions in digester operations.

4.0 SYSTEMS EVALUATIONS OF FEEDSTOCK PREPROCESSING

4.1 Process Integration of Feedstock Preprocessing Equipment

The integration of preprocessing equipment is determined by the feedstock types, contaminations tolerance of the digestion system, and intended scope of the co-digestion program. The co-digestion and preprocessing of food waste offer a multifaceted solution to some of the pressing environmental challenges associated with waste management. Figure 5.8 shows a comparison of suitable mechanical decontamination technology based on contaminant type based on interview feedback with a higher score indicating a more favorable application. Table 5.5 listed three process flow schematics based on full-scale installations. Due to the changes of feedstock availability and tipping fee volatility, reconfigurable SSO preprocessing (Wang et al., 2018), as inspired by smart manufacturing (Huang et al., 2022), attracts more attention in innovative process design and technology development. Future research and technological advancements in this field hold the potential to further refine these processes, making them more efficient, cost-effective, and widely applicable.

The preprocessing techniques—physical, chemical, and biological—each play a critical role in optimizing the co-digestion process. Physical

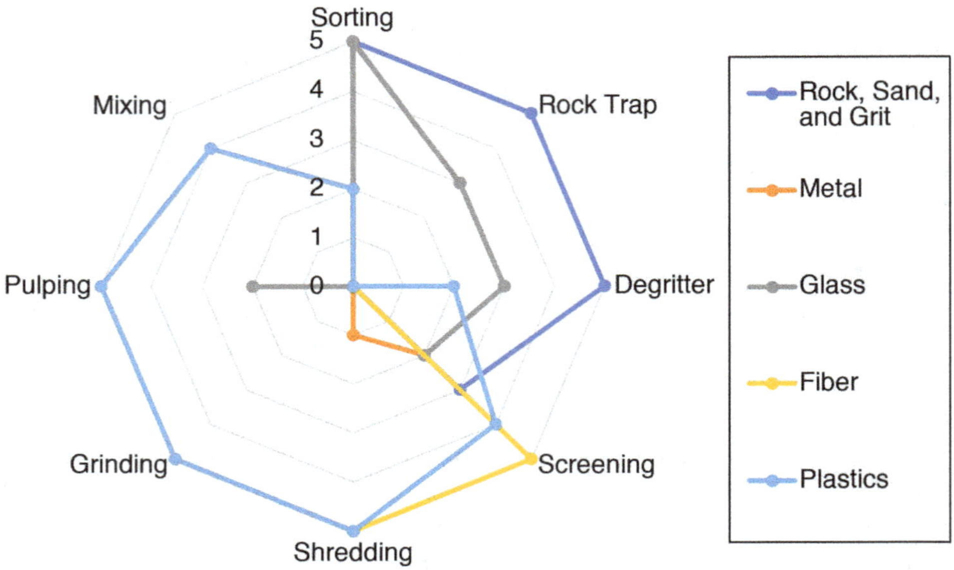

FIGURE 5.8 Suitable Decontamination Technology by Contaminant Type

preprocessing methods, such as grinding and screening, help in reducing particle size and removing contaminants, making the waste more homogenous. Chemical preprocessing techniques, including pH adjustment and the use of additives, can further aid the biochemical process where complex organic materials are broken down, enhancing their digestibility. Biological preprocessing methods, such as enzymatic treatments and fermentation, can pre-digest the waste, increasing the efficiency of the subsequent co-digestion process. By integrating these preprocessing techniques, the overall efficiency of the co-digestion process can be significantly improved, leading to higher yields of biogas and other valuable by-products.

The configuration of the feedstock preprocessing is site and equipment specific. System integration plays a key role in successfully implementing SSO preprocessing and co-digestion. The discrete nature of raw organic waste from pre-consumer and post-consumer sources implies the strategies of optimizing sustainable manufacturing processes could be beneficially applied for the design and optimization of SSO preprocessing based on specific need at a facility (Chowdhury et al., 2021; Jin et al., 2019). For instance, the Los Angeles County Sanitation Districts (LACSD) receive unprocessed food waste at the solids waste management facility, where a customized food waste preprocessing and holding station converts raw food waste to preprocess bioslurry (Figure 5.9). The LACSD also accepts third-party preprocessed food waste at the A. K. Warren Water Resource Facility through odor-controlled receiving stations with minimal additional

TABLE 5.5 Process Flow Schematics of SSO Preprocessing Facility

Process Flow Schematics	Typical Application Scenarios
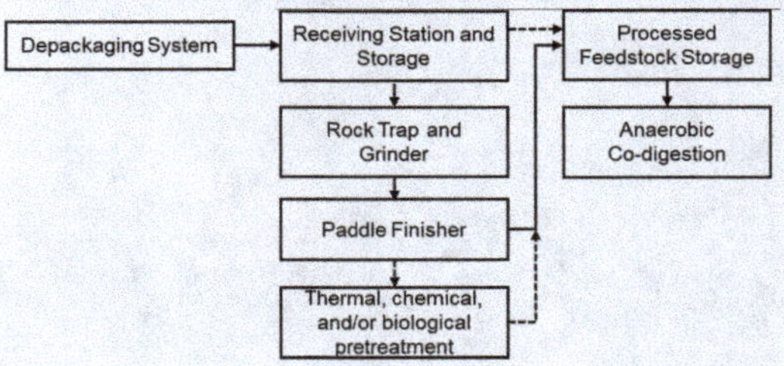	• Feedstock consists of pre-consumer SSO waste such as beverages, packed meat, and processed food. • Separation preprocessing such as rock trap, grinder, and paddle finisher could be bypassed if contamination level in feedstock is low. • Optional thermal, chemical, and biological pretreatment.
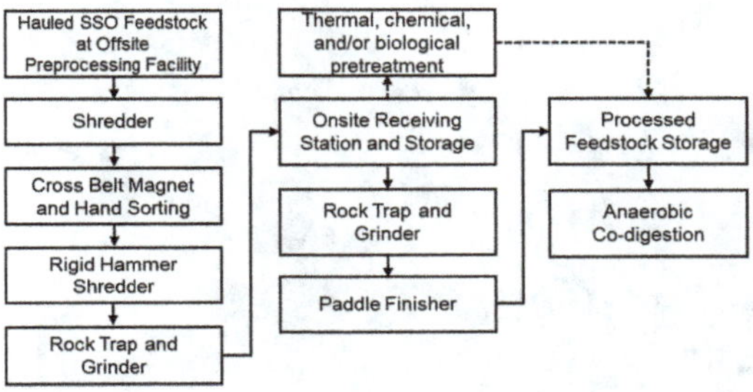	• A combination of offsite and onsite preprocessing. • Capable of receiving diverse sources of feedstock, including SSO derived from post-consumer residential organic waste. • Optional thermal, chemical, and biological pretreatment. • Complex process may require dedicated O&M staff.
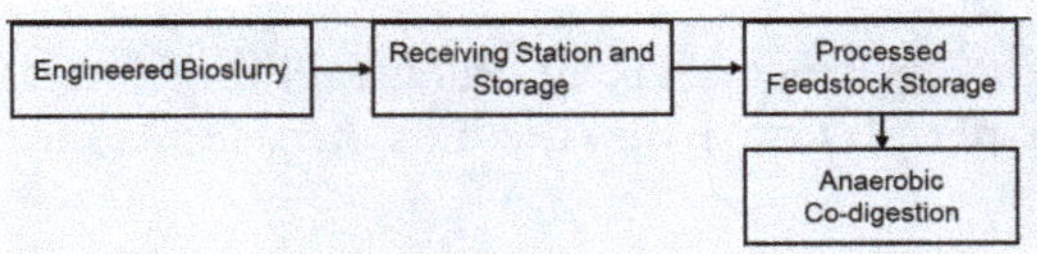	• Feedstock is restricted to engineered bioslurry. • Minimal on-site preprocessing is required. • Feedstock quality can be customized based on site specific requirements.

preprocessing (Figure 5.10). The City of Muscatine Water and Resource Recovery Facility, as an example of medium-sized WRRF, converted a portion of existing building (with high clearance) for food waste recycling and preprocessing (Figure 5.11).

4.2 Monitoring and Evaluation Approaches of Feedstock Preprocessing Processes

System-level considerations of monitoring tools and sensors, equipment reliability, process performance, and impacts on the compliance of biosolids management and liquid stream effluent discharge. A structured evaluation will guide facility planning effort in evaluating the technology feasibility of the target SSO feedstock market, identifying the operation and maintenance (O&M) capability, and assessing the potential impact on normal operation

FIGURE 5.9 Food Waste Preprocessing Process at the Puente Hills Material Recovery Facility (Reprinted with permission from the Los Angeles County Sanitation Districts)

of the hosting WRRF. A system-level evaluation approach is proposed in Figure 5.12.

Techno-economic performance of feedstock preprocessing facility has a profound impact on the sustainability of co-digestion programs. Surveys show that tipping fee revenue and boost of biogas production are most important economic drivers for co-digestion program (Jones et al., 2019). Unlike typical wastewater treatment or biosolids handling processed for which the flows and loads are predictable based on historical data of the existing WRRF or facility of similar size, the flows, loads, and contamination levels vary by the types of SSO feedstock, market condition of the waste recycling industry, and the production schedule of major SSO generators such as food and beverage production facilities. In addition, the funding sources for co-digestion programs are more dynamic than typical wastewater infrastructure that usually involves more stakeholders' involvement for funding justification. A comprehensive techno-economic analysis (TEA) may provide a granular evaluation of multiple technology routes by incorporating key variables through faction factorial design. A comprehensive TEA

FIGURE 5.10 Preprocessed Food Waste Slurry Receiving and Holding Stations at the A. K. Warren Water Resource Facility (Reprinted with permission from the Los Angeles County Sanitation Districts)

FIGURE 5.11 Food Waste Receiving and Preprocessing (Reprinted with permission from the City of Muscatine Water and Resource Recovery Facility)

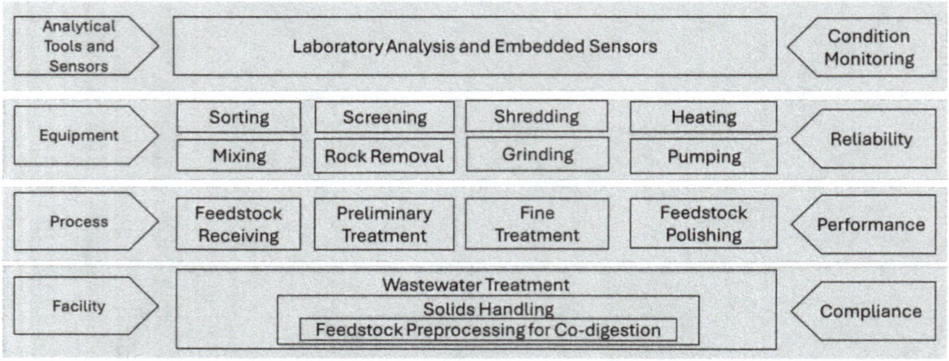

FIGURE 5.12 System-level Evaluation of Feedstock Preprocessing

will also provide guidance on the feasibility of system scale-up, especially for utilities searching federal and state assistance in adopting innovative processes in WRRFs. Figure 5.13 shows a TEA roadmap that is adopted from the U.S. Department of Energy Technology Readiness Assessment Guide (U.S. Department of Energy, 2015) and have been applied for sustainable aviation fuel production through biomass (Shakelly et al., 2024).

Co-digestion preprocessing systems are impacted by degradation and failures caused by the harsh operation conditions. Maintenance strategies aim to improve the availability of the preprocessing system. However, maintenance represents considerable cost in terms of resources, time, tools, and spare part inventories. Maintenance costs can vary between 15% to 70% of the total operation cost. Preventive maintenance is a common strategy

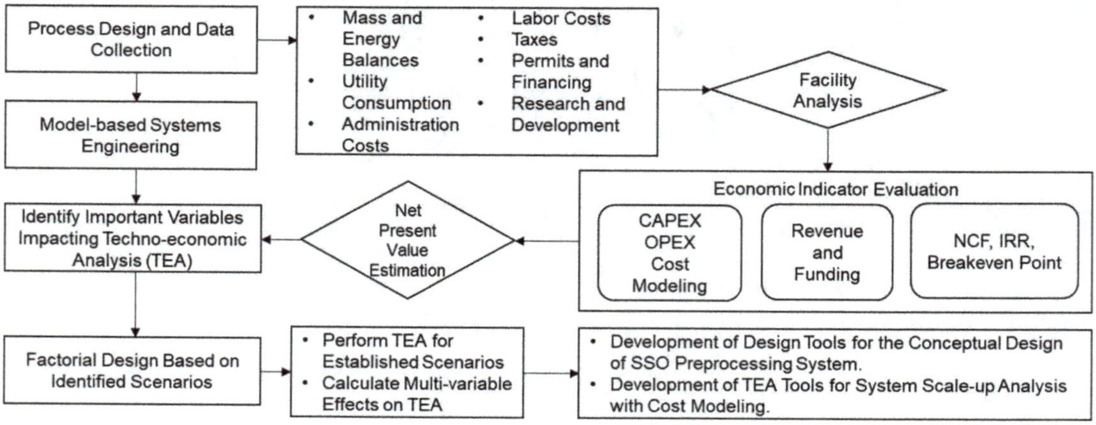

FIGURE 5.13 Techno-economic Analysis Roadmap for Sanitary Sewer Overflow Feedstock Preprocessing Evaluation

used to guarantee the availability and efficiency of production systems (Arismendy et al., 2020).

Predictive maintenance (PdM) is a condition-based maintenance strategy that conduct maintenance action when needed based on the state of the machine, avoiding unnecessary preventive actions and premature failures (Wu et al., 2023). Predictive maintenance uses predictive models based on condition monitoring to ensure the availability of feedstock preprocessing equipment by assessing the machine's state. Predictive maintenance also provides early warning signals when failure is incipient (Joung et al., 2024). As such, PdM allows production time optimization, maintenance scheduling, spare part inventory planning, and real-time receiver operating characteristic area under the curve-evaluation. As shown in Figure 5.14, key elements of PdM include (1) define the operation model of the component or a machine tool and its failure mode to establish the predictive goal of PdM; (2) deploy sensor networks for data collection; (3) conduct data mining through data selection, analysis, processing, modeling, and evaluation for validating the predictive model; and (4) implement a decision-making framework and recommend proper maintenance action. It is recommended to reference Chapter 7, "Co-digester Operations and Maintenance" for additional information.

5.0 SUMMARY

The management of food waste through co-digestion preprocessing presents a promising avenue for enhancing environmental sustainability and resource

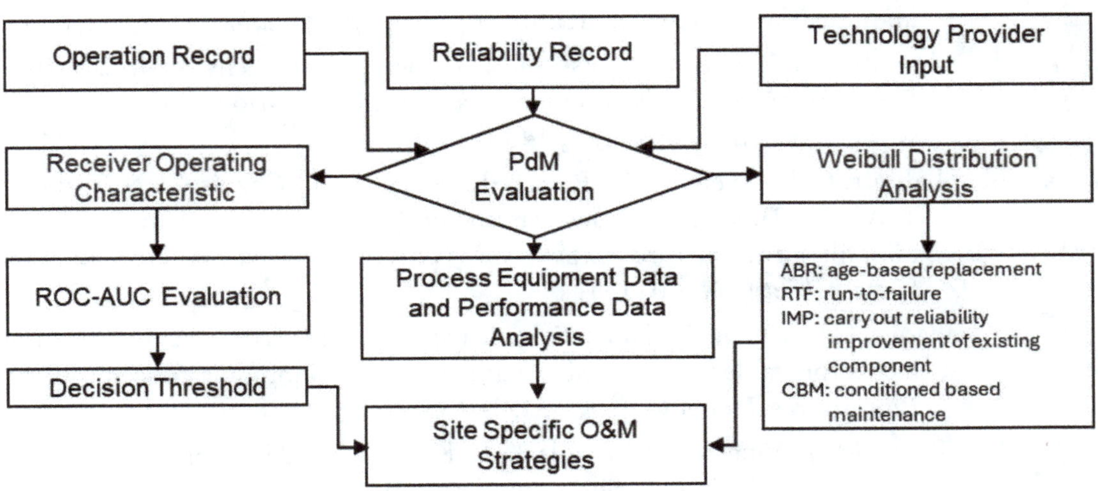

FIGURE 5.14 Site Specific Operation and Maintenance Strategies for Sanitary Sewer Overflow Feedstock Preprocessing

efficiency. By understanding the complexities of food waste, including its inconsistent composition, high contaminant levels, and logistical challenges, stakeholders can better appreciate the need for effective preprocessing techniques. Given the fact that only a small fraction of WRRFs have adopted food waste co-digestion and the rapid development of SSO preprocessing technologies, methodology development is required to evaluate the system-level impact of SSO preprocessing on co-digestion, biosolids management, and broader compliance impact for WRRFs. The tailored techno-economic analysis will guide decision-makers in understanding the dynamic interaction of facility size and return on investment of co-digestion programs, as well as the roles preprocessing plays in the planning phase of a project. A practical and robust operation and maintenance plan is essential for scheduling equipment to accommodate a facility-wide energy management plan for in-demand bioenergy generation at WRRFs.

6.0 REFERENCES

Abu-Orf, M., Dursun, D., Blate, M., Landry, K., Murray, C., Sharp R., & Kaiser, N. (2023). *Evaluation of source separated organic feedstock pretreatment and management practices*. Water Research Foundation.

Almansa, A. R., Rodriguez-Galan, M., Borja, R., & Fermoso, F. G. (2015). Micronutrient dynamics after thermal pretreatment of olive mill solid waste. *Bioresource Technology, 191*, 337–341.

Arismendy, L., Cárdenas, C., Gómez, D., Maturana, A., Mejía R., & Quintero M., C. G. (2020). Intelligent system for the predictive analysis of an industrial wastewater treatment process. *Sustainability*, *12*(16), 6348.

Ariunbaatar, J., Panico, A., Esposito, G., Pirozzi F., & Lens, P. N. L. (2014). Pretreatment methods to enhance anaerobic digestion of organic solid waste. *Applied Energy*, *123*, 143–156.

Babbitt, C. W., Neff, R. A., Roe, B. E., Siddiqui, S., Chavis C., & Trabold, T. A. (2022). Transforming wasted food will require systemic and sustainable infrastructure innovations. *Current Opinion in Environmental Sustainability*, *54*, 101151.

Banks, C. J., Zhang, Y., Jiang Y., & Heaven, S. (2012). Trace element requirements for stable food waste digestion at elevated ammonia concentrations. *Bioresource Technology*, *104*, 127–135.

Bardi, M. J., Vinardell, S., Astals S., & Koch, K. (2023). Opportunities and challenges of micronutrients supplementation and its bioavailability in anaerobic digestion: A critical review. *Renewable and Sustainable Energy Reviews, 186*, 113689.

Bong, C. P. C., Lim, L. Y., Lee, C. T., Klemeš, J. J., Ho, C. S., & Ho, W. S. (2018). The characterisation and treatment of food waste for improvement of biogas production during anaerobic digestion—A review. *Journal of Cleaner Production*, *172*, 1545–1558.

Carol, A. J., Coker, C., & Kirk, K. (2019). *Food waste co-digestion at water resource recovery facilities: Business case analysis*. The Water Research Foundation.

Centemero, M. R., & Confalonieri, A. (2024). Italy's experience with compostable plastics in organics recycling. https://www.biocycle.net/italy-compostable-plastics/

Cheng, H., Li, Y., Guo, G., Zhang, T., Qin, Y., Hao, T., & Li, Y. Y. (2020). Advanced methanogenic performance and fouling mechanism investigation of a high-solid anaerobic membrane bioreactor (AnMBR) for the co-digestion of food waste and sewage sludge. *Water Research*, *187*, 116436.

Chowdhury, N. A., Deng, S., Jin, H., Prodius, D., Sutherland, J. W., & Nlebedim, I. C. (2021). Sustainable recycling of rare-earth elements from ndfeb magnet swarf: Techno-economic and environmental perspectives. *ACS Sustainable Chemistry & Engineering*, *9*(47), 15915–15924.

Coker, C. (2021). Food depackaging: The systems. *BioCycle*. https://www.biocycle.net/food-depackaging-the-systems/

Elbeshbishy, E., & Nakhla, G. (2011). Comparative study of the effect of ultrasonication on the anaerobic biodegradability of food waste in single and two-stage systems. *Bioresource Technology*, *102*(11), 6449–6457.

Huang, A., Triebe, M., Li, Z., Wu, H., Joung, B. G., & Sutherland, J. W. (2022). A review of research on smart manufacturing in support of environmental sustainability. *International Journal of Sustainable Manufacturing*, *5*(2–4), 132–163.

Iacovidou, E., Ohandja, D.-G., & Voulvoulis, N. (2012). Food waste co-digestion with sewage sludge—Realising its potential in the UK. *Journal of Environmental Management*, *112*, 267–274.

Jain, S., Jain, S., Wolf, I. T., Lee J., & Tong, Y. W. (2015). A comprehensive review on operating parameters and different pretreatment methodologies for anaerobic digestion of municipal solid waste. *Renewable and Sustainable Energy Reviews*, *52*, 142–154.

Jin, H., Reed, D. W., Thompson, V. S., Fujita, Y., Jiao, Y., Crain-Zamora, M., Fisher, J., Scalzone, K., Griffel, M., Hartley, D., & Sutherland, J. W. (2019). Sustainable bioleaching of rare earth elements from industrial waste materials using agricultural wastes. *ACS Sustainable Chemistry & Engineering*, *7*(18), 15311–15319.

Jones, C. A., Coker, C., Kirk K., & Reynolds, L. (2019). *Food waste co-digestion at water resource recovery facilities: Business case analysis.* Water Research Foundation.

Joung, B. G., Nath, C., Li, Z., & Sutherland, J. W. (2024). Bearing anomaly detection in an air compressor using an LSTM and RNN-based machine learning model. *The International Journal of Advanced Manufacturing Technology*, *134*(7), 3519–3530.

Kovacic, D., Radocaj, D., & Jurisic, M. (2024). Ensemble machine learning prediction of anaerobic co-digestion of manure and thermally pretreated harvest residues. *Bioresource Technology*, *402*, 130793.

Mabe, L., Pace, S. A., & Spang, E. S. (2022). A cluster-based spatial analysis of recycling boundaries aligning anaerobic digestion infrastructure with food waste generation in California. *Resources, Conservation & Recycling Advances*, *15*, 200113.

Nghiem, L. D., Koch, K., Bolzonella, D., & Drewes, J. E. (2017). Full scale co-digestion of wastewater sludge and food waste: Bottlenecks and possibilities. *Renewable and Sustainable Energy Reviews*, *72*, 354–362.

Pognani, M., Barrena, R., Font, X., Scaglia, B., Adani F., & Sánchez, A. (2010). Monitoring the organic matter properties in a combined anaerobic/aerobic full-scale municipal source-separated waste treatment plant. *Bioresource Technology*, *101*(17), 6873–6877.

Rauch-Williams, T., Raskin, L., Schaum, C., & Gupta, R. (2023). *Characterization of source separated food waste for co-digestion in water resource recovery facilities*. Water Research Foundation.

Sagar, N. A., Pathak, M., Sati, H., Agarwal, S., & Pareek, S. (2024). Advances in pretreatment methods for the upcycling of food waste: A sustainable approach. *Trends in Food Science & Technology*, *147*: 104413.

Shakelly, N., Li, Z., Deng, S., & Sutherland, J. W. (2024). Effect of key economic uncertainties on the techno-economic performance of the production of high energy-content jet fuel blending components derived from bio-ethanol. *Sustainable Energy Technologies and Assessments*, *65*, 103747.

Smith, P. G. (2005). *Dictionary of water and waste management.* Butterworth-Heinemann.

Szmidt, R. A. K. (2022). Pretreatment equipment and food waste preparation for anaerobic digestion. *Food Waste Valorisation*, WORLD SCIENTIFIC (EUROPE), 149–184.

U.S. Department of Energy. (2015). *Technology readiness assessment guide.*

U.S. Environmental Protection Agency. (2021). *Anaerobic digestion facilities processing food waste in the U.S. (2020 & 2021).*

U.S. Environmental Protection Agency. (2022). *Downstream management of organic waste in the United States: Strategies for methane mitigation.*

Usmani, Z., Sharma, M., Awasthi, A. K., Sharma, G. D., Cysneiros, D., Nayak, S. C., Thakur, V. K., Naidu, R., Pandey A., & Gupta, V. K. (2021). Minimizing hazardous impact of food waste in a circular economy—Advances in resource recovery through green strategies. *Journal of Hazardous Materials*, *416*, 126154.

Wang, Y., Mendis, G. P., Peng S., & Sutherland, J. W. (2018). Component-oriented reassembly in remanufacturing systems: Managing uncertainty and satisfying customer needs. *Journal of Manufacturing Science and Engineering, 141*(2).

Wu, H., Triebe M. J., & Sutherland, J. W. (2023). A transformer-based approach for novel fault detection and fault classification/diagnosis in manufacturing: A rotary system application. *Journal of Manufacturing Systems*, *67*, 439–452.

Zhang, S., Liang, C., Xiao, M., Chui, C., Wang, N., Ji, Y., Wang, Z., Shi, J., & Liu, L. (2023). Metagenomic characterization of the enhanced performance of multicomponent synergistic thermophilic anaerobic co-digestion of food waste utilizing kitchen waste or garden waste as co-substrate. *Water Research*, *244*, 120457.

Zitomer, D. H., Adhikari, P., Heisel C., & Dineen, D. (2008). Municipal anaerobic digesters for codigestion, energy recovery, and greenhouse gas reductions. *Water Environment Research*, *80*(3), 229–237.

Chapter 6

System Design Considerations

Christopher Buckley, PE; Jason Rafferty, PE; & Amy Underwood, PE

1.0 REGULATORY AND ENVIRONMENTAL REQUIREMENTS

1.1 National Pollutant Discharge Elimination System Regulations

Co-digestion design and operations are not typically regulated by the water resource recovery facility's (WRRF's) National Pollutant Discharge Elimination System (NPDES) or sludge hauling/disposal permits per se. However, a WRRF is expected to continue to meet all its NPDES permit regulations. This includes WRRF site stormwater NPDES requirements.

1.2 Capacity Management, Operations, and Maintenance

Co-digestion operations require substantially more effort in operations and management. Provisions must be made for managing the co-digestion process as well as feedstock receiving, testing, and storage.

Co-digestion requires significantly more sampling and monitoring of the digestion process than conventional anaerobic digestion. Anaerobic digestion involves complex chemical and biological processes that can be easily upset by frequent changes to the quality and quantity of feedstocks. Frequent and accurate monitoring of critical digestion process parameters such as volatile acids, pH, and alkalinity are essential to ensure satisfactory operation of the digesters under co-digestion.

The additional maintenance required for feedstock acceptance, storage, and co-digestion should not be underestimated. Most feedstocks can have a viscous characteristic that increases cleaning requirements. It has been frequently observed that cleaning receiving stations and clearing clogged pipes, pumps, valves, and tanks requires a substantial amount of time and effort by publicly owned treatment works (POTW) maintenance staff.

Such operations and maintenance (O&M) needs must be clearly identified, discussed, and planned before initiating co-digestion operations. Neglecting to prepare for such O&M requirements is reported to be the leading cause of POTW co-digestion projects failing or being underutilized.

1.3 Design Approvals

Regulations vary by state and locality, but co-digestion design and operations are generally not regulated for POTWs other than with broad restrictions. Most POTW design regulations focus on raw sludge derived from "ordinary domestic wastewater" (Great Lakes—Upper Mississippi River Board of State and Provincial Public Health and Environmental Managers, 2014). Digestion of organic feedstocks has traditionally been regulated as an industrial and/or wastewater process, typically a pretreatment step before discharge to the POTW.

For instance, 10 States Standards does not mention co-digestion (other than primary sludge and waste activated sludge [WAS] co-digestion) except to avoid high concentrations of sulfates and inhibitory concentrations of heavy metals (Great Lakes—Upper Mississippi River Board of State and Provincial Public Health and Environmental Managers, 2014). Most anaerobic digestion design guidelines rely on the design engineer providing a rational design basis that involves a comprehensive system design, including tank volumes, kinetic calculations, industry design information, and current operating records.

Some states (such as Illinois) allow for an "empirical design" that limits the organic loading of digesters based on tank capacity (kg of volatile suspended solids [VSS]/28.32 m^3 [lb of VSS/1000 CF] per day) and the type of mixing used (Illinois Pollution Control Board, 2024). As mentioned, these regulations are based on "raw sludge derived from ordinary domestic wastewater," and deviations (such as co-digestion) must demonstrate that the existing digestion capacity is adequate or physically provide extra capacity.

Along those lines, co-digestion pilot studies are highly recommended before initiating a full-scale co-digestion design. A pilot study can help assess the impact on the existing digestion system and demonstrate the operations and maintenance requirements. It also provides data supporting a rational design for permitting. Since there are not many regulations for co-digestion, regulatory agencies have an easier time issuing construction permits for co-digestion if there is demonstrated experience.

Ultimately, a co-digestion design must, at minimum, avoid degrading of POTW effluent quality and violating NPDES permit requirements. Additionally, co-digestion cannot generate substantial additional odors. Lastly, the classification/quality of current biosolids reuse/disposal should be maintained. However, a change in biosolids reuse/disposal classification could be part of an overall co-digestion master plan.

For co-digestion designs that involve new facilities, local regulations may require the submittal of design plans to the local building authority for review. Typically, these reviews are more concerned about local impacts, building codes, and environmental issues than the co-digestion process. However, one issue with co-digestion and feedstock acceptance that may be outside the typical environmental protection agency (EPA) state regulatory review process is the substantial additional truck traffic to the POTW that may impact neighbors and local traffic.

1.4 Environmental Issues and Compliance

The main environmental issues with co-digestion are the POTW's continuing adherence to its NPDES, sludge reuse/disposal requirements, and problems that occur "outside the fence." These exterior problems may include:

1. Local traffic and roadway impacts: Feedstock hauling to the WRRF could increase truck traffic by a considerable amount. Large-scale feedstock hauling may use vehicles that can damage local roadways. A study of the impacts of feedstock hauling operations should be carefully considered.
2. Odors: Odor control measures may need to be considered for POTWs. The type and magnitude of odor generated from co-digestion can cause additional complaints. Publicly owned treatment works

occasionally have odor nuisances, which the public typically tolerates. However, because co-digestion is not a regulatory-required process, odors from feedstocks and poor co-digestion may be met with public hostility. In some cases, co-digestion operations have been suspended due to public pressure.

Feedstock receiving and storage can be a source of odors but generally can be controlled with sound operations, frequent facility cleaning, and passive odor control. One of the more serious odor problems can occur during digester process upsets. The "souring" of a digester, especially one co-digesting feedstock, can generate strong, offensive odors that last for weeks. Such occurrences can be minimized with careful operations and monitoring of digester health.

3. Stormwater runoff: Most POTWs have stormwater runoff restrictions (Stormwater Pollution Prevention Plans) requiring POTWs to control stormwater runoff from carrying potential pollutants off the POTW site. These plans must be reviewed and potentially revised regarding feedstock handling and storage. It is best to design feedstock receiving facilities to collect any spillage or runoff and direct it back to the head of the POTW for containment and treatment.

2.0 DESIGN GUIDELINES

2.1 Design Process Checklist

The following is a list of required information for planning and designing a co-digestion program.

2.1.1 Design Plans and Bases of Designs

Obtain existing design plans and basis of design reports on the existing digestion facilities as well as the WRRF site.

2.2.1 Plant Operating Records

Obtain a minimum of 2 years and preferably 3 years of existing facility operating data, primarily regarding the existing anaerobic digestion process:

1. Influent sludge (PS and/or WAS flows, biochemical oxygen demand/chemical oxygen demand [BOD/COD], total suspended solids [TSS], VSS, pH, sulfates)
2. Internal digester process:
    a. Volatile acids
    b. pH

   c. Sludge retention time/hydraulic retention time (SRT/HRT)
   d. Alkalinity
   e. Sulfate
3. Digested sludge (flows, BOD/COD, TSS, VSS, pH, sulfates)
4. Gas production (flows, pressure, temperature, methane, hydrogen sulfide, carbon dioxide, and siloxane concentrations).

2.1.3 Feedstock Information

A review of the type and quantity of feedstocks that might be initially utilized should be obtained (see Chapters 3, 4, and 5). Information to be received includes detailed feedstock-related physical and chemical descriptions, laboratory sampling, estimated quantities, frequency of deliveries, transportation types, and unloading options. Any chain of custody procedures, transfer of liability requirements, or minimum guaranteed quantities requirements should also be obtained.

Unless within a major metropolitan area, utilizing available feedstocks may be limited to the local industries near the WRRF area, as transportation costs for hauling feedstocks long distances may be cost-prohibitive. A common genesis of WRRF co-digestion facilities occurs from the availability of one or two local sources of feedstock seeking to dispose of their organic waste more cost-effectively.

A review of other potential feedstocks in the area should be conducted to ascertain the potential feedstock market and diversify feedstock sources. However, it is not recommended to utilize more than one or two different feedstocks at a time, as this can cause upsets in the co-digestion process.

2.2 Digestion Capacity Assessment and Modifications

To utilize existing anaerobic digesters, or install new ones, for co-digestion, several factors require careful analysis. These include digester solids loading to verify that the existing digester can as well as have the approved any local regulatory agencies. Additionally, the WRRF and existing digesters may need to be physically modified to effectively co-digest. Failure to address these issues can result in failure of a co-digestion project.

2.2.1 Digester Solids Loading

To assess the efficacy of co-digestion, the capacity of existing anaerobic digesters (or capacity required for new digesters) needs to be assessed from a regulatory perspective to verify what organic limitations may limit co-digestion utilization. Once established, the design can be based on as well reviewing existing digester operating data.

2.2.1.1 Regulatory Limitations

Some states regulate the maximum organic capacity of a digester. As mentioned, Illinois has a standard digester organic loading of 1.27 kg VSS/m^3 (80 lb VSS/1000 CF) per day based on providing mechanical mixing (see more on this below).

It should be noted that these are design standards that pertain to providing sufficient digester capacity based on the WRRF's organic loading. These standards are not necessarily operating limitations. It is common for WRRFs to operate digesters successfully at much higher loading rates.

However, if state EPA agencies require the submission of construction documents for any new WRRF facilities, regulators may review the project against these design standards.

2.2.1.2 Co-digestion Facility Operating Data

To assist facilities with assessing their potential co-digestion capacity, the following is a list of WRRFs currently co-digesting. Included in Table 6.1 is a comparison of their design digesting loading rates versus current loading rates with the addition of feedstocks. Loadings include both process sludge and feedstock.

Per Table 6.1, anaerobic digesters have been loaded with feedstock 25% to 50% above conventional design capacities without upset. However, these loading rates are achieved over time, with the digesters carefully fed with quality feedstocks and monitored closely.

The percentage of feedstocks to process sludge varies based on each system. For instance, DGSD (as seen in Table 6.1) has been operating at a feedstock fats, oil, and grease (FOG) VSS loading rate of 67% of their total digester organic loading. However, they have been experiencing a reduction in biosolids reuse/disposal quality (described later in this chapter) and are contemplating limiting their feedstock loading rate to only 40% of their total digester loading to regain biosolids reuse/disposal quality (Downer's Grove Sanitary District, 2024).

2.2.2 Digester Modifications

Most co-digestion projects attempt to utilize their existing digester facilities. This approach is generally sound, but several key digester modifications may be required.

2.2.2.1 Mixing

Mixing digesters is critical to achieving thorough co-digestion and reducing operational difficulties.The main objective of mixing is to improve contact

TABLE 6.1 Existing Water Resource Recovery Facility Co-digestion Facility Operating Data

FACILITY	Digester Capacity m^3 (CF)	Average Organic Co-digestion Loading Rate kg VSS/m^3/day (Lb VSS/kcf/day)	Maximum Sustained Organic Co-digestion Loading Rate kg VSS/m^3/day (Lb VSS/kcf/day)	Feedstock % of Total VSS Mass Loading Rate %	Remarks
Downers Grove Sanitary District (DGSD), IL*	3842 (135 695)	1.16 (72)	1.52 (95)	67%	Co-digestion with WAS. Separate primary sludge digesters.
Glenbard Wastewater Authority (GWA), IL	4278 (151 069)	1.6 (100)	2.66 (166)	35%	Co-digestion with PS and WAS.
Kishwaukee Water Reclamation District (KWRD), IL*	(3842) (135 695)	1.44 (90)	1.6 (100)	40%	Co-digestion with WAS. Separate primary sludge digesters.
Sheboygan, WI					

Active mixed and heated digester capacity

between microorganisms and substrates. It's important to note that in this text, "digester mixing" refers to the ability to mix the digester using methods other than sludge influent/withdrawal pumping or sludge heating recirculation pumping. Those systems are primarily designed to move sludge into and out of the digester but do not adequately mix the digester contents (Sludge Process and Treatment, EPA 625/1).

Inadequate mixing can cause poor digestion as the digester contents tend to stratify, thus limiting the organics' exposure to the microorganisms and heat that facilitate decomposition and methane gas generation. Eventually, the digester can lose substantial capacity as the unmixed digester contents can stratify and settle over time.

As noted, some regulatory guidelines limit the digester capacity based on the type of mixing provided. High-rate capacity digesters are only permitted if they use mechanical mixing rather than convection or passive mixing. For co-digestion projects, mechanical mixing is almost always required to maximize the capacity of the digester and achieve the best VSS destruction as well as the highest gas generation rates. Mixing is especially critical with co-digestion as typical feedstock can exhibit viscous behavior (non-Newtonian fluid) that inhibits mixing.

Key tank mixing design parameters involve some general design parameters:

1. Mixing power: Power inputs range from 0.15 to 0.22 kW/28.32 m^3 (0.2 to 0.3 hp /1000 CF) of digester capacity (U.S. EPA, 1979).
2. Velocity gradient: A method of measuring mixing intensity measured in units per time. The higher the gradient, the easier it is to mix the liquid. Conventional digester contents have a typical velocity gradient of 50 to 80 sec-1 (Schlicht, 1999). A more viscous liquid has a lower velocity gradient and is thus more challenging to mix.
3. Turnover time: Time required to recirculate the entire digester's contents. It is calculated as the volume of the digester divided by the mixing device's flow or liquid displacement rate. The higher the rate, the more the digester is mixed. However, this assumes that prescribed amount of fluid that is actually being moved through the digester (Schlicht, 1999).

These are long-standing design guidelines that attempt to quantify how much mixing hp power is required to mix the entire digester contents efficiently. However, other factors, such as tank geometry and mixer design, can impact mixing effectiveness. Mixing anaerobic digesters is challenging due to the varying fluid physical parameters in the tank as well as the fact that it is difficult to measure mixing effectiveness in such an enclosed environment.

Fortunately, in the last 20 years, mixer manufacturers have started using sophisticated analysis tools such as tracer tests and computational fluid dynamics (CFD) to model mixing performance more effectively. Any analysis of digester mixing requirements should involve an equipment manufacturer doing a thorough analysis of the tank with accurate information available on tank geometry and fluid characteristics.

Another major factor impacting mixing (and co-digestion in general) is debris within the digester can decrease mixing performance and reduce capacity. Sand, grit, hair, rags, and other debris negatively impact the mixing efficiency. Sludge grinders can help improve mixing but reduce the debris into smaller pieces. Finer screening may be required upstream in the WRRF process to achieve maximum digestion capacity.

There are several ways to input the power required to turn over the digester and maximize co-digestion. Typical digester mixing designs include:

- Gas mixing
- Pumped mixing
- Mechanical mixing
- Static mixing
- Momentum mixing

Gas mixing: Generally, gas (biogas) mixing utilizes gas compressors to circulate produced and stored digester gas throughout the digester to mix the contents. Biogas is injected into the bottom of the digester through eductors, spargers, or "bubble guns" and it percolates to the top of the digester, mixing the digester contents in between via uplift force, turbulence, or both. Biogas is then collected at the top of the digester to continue the process.

Gas mixing has the advantage of being a relatively simple process with low capital cost. Effective mixing has been reported, but results are varied. Gas mixing can create localized mixing "holes" around the spargers or along the wall edges, which does not fully mix the digester. Gas mixing has also been reported to have difficulty with mixing floatables like scum and FOG. Gas mixing has become less utilized in the last few decades due to its high operations and maintenance requirements, compressed gas safety concerns, and the shift to using digester gas for waste recovery purposes rather than mixing.

Pumped mixing: Pumped mixing involves using high-flow, non-clog recirculation pumps outside the digester that connects to a system of specially designed nozzles inside the digester. This system generates high-velocity flows within the digester, creating an upward-to-downward, circular mixing pattern.

As noted, pumped mixing is separate from sludge influent/withdrawal pumping or sludge heating recirculation pumping. Commonly provided sludge heating recirculation systems through boilers/heat exchangers often do not fully mix the digester contents.

Pumped mixing has become more commonplace in the last few decades through the use of dynamics CFD to model the most effective and cost-efficient nozzle locations while minimizing the pump flows and motor horsepower required. Sludge mixing pumps have also become more reliable with improved impeller designs, such as chopper type and screw centrifugal type, that pump various digester contents without frequent clogging. Chopper pumps are a type of non-clog pump with a sharpened impeller edge that promotes slicing up solids as they are pumped. Screw centrifugal pumps offer a wide passage impeller that helps to pump large-diameter solids. These pumps have slightly lower efficiencies than a standard non-clog but have effectively reduced clogging and improved mixing in anaerobic digester applications.

Pumped mixing is reported to have many successful installations that create a fully mixed digester. Long-term operations at several facilities have noted that good mixing (based on empirical data such as increased volatile solids destruction) can often be obtained by having several strategically timed short mixing periods versus continuous mixing. However, pumped mixing does have high capital and energy costs and can have foaming issues.

Mechanical mixing: Mechanical mixing, also known as draft tube mixing, involves installing a propeller type (axial) pump on the top of a pipe leading into the digester. The pump impeller sits in the pipe as an extended shaft and induces a suction (draft) pulling sludge into the pump. Variations in the design include being mounted on top of the digester or along the sides. Advantages include high flows, reversible flow, and less foaming. Disadvantages include high capital costs, high maintenance costs to remove the pump for servicing as well as poorer mixing if the suction piping gets clogged. Also, draft tubing mixing can have problems with heat loss as well and freezing in northern climates. On the other hand, some installations have installed heat exchangers on the external draft tubes to provide sludge heating rather than providing internal heat exchangers.

Static mixers: Static mixers, or vertical agitator mixers, consist of a large propeller mixer suspended within the digester tank that is supported on top of the cover. Mixing is promoted via a churning action that forces the digester contents upward and downward against the tank structure, creating upward flow circulation. Advantages include simplified installation with low power consumption. Disadvantages are high maintenance costs, especially to remove and service the mixer, problems with ragging around the propeller shaft, and potential inadequate mixing with digesters that have large amounts of grit or debris.

Momentum mixing: A newer digester mixing technology, momentum mixing, uses a large plunger to generate the mixing energy within the digester via a top-mounted exterior reciprocating drive. The method has gained traction as a more straightforward, less energy-intensive method of mixing digesters than gas or pumped mixing. However, maintenance issues have been reported related to the frequent replacement of drive equipment.

It is recommended to conduct a thorough review of a site's current mixing design when considering co-digestion to ascertain if the digester mixing system may need to be upgraded. The choice of mixing technology is very site-specific and depends on the co-digestion site location, size of digesters, and the WRRF O&M capabilities. It is advisable to give equipment suppliers as much information about potential co-digestion and feedstock physical characteristics so they may take that into account when advising on equipment selection.

2.2.2.2 Heating

For this publication, co-digestion is assumed to occur using mesophilic digestion, which involves heating the digester contents to an average temperature of 95 °F (35 °C) for a minimum of 15 days of solids retention time (U.S. EPA, 1979). External heat exchangers with gas-fired boilers are often used to heat incoming "cold" sludge and recirculate it within the digester to maintain the reactor temperature. Frequent use of the additional biogas produced by co-digestion is heating for these boilers, thus creating a sustainable heating system.

Reports on co-digestion's additional thermal energy requirements are varied. Some POTWs have observed no net change in heating requirements (according to Matt Streicher of Glenbard Wastewater Authority, 2024). However, one POTW reported increasing its boiler/heat exchanger set-point temperatures to 98° during wintertime after commencing co-digestion operations. This suggests that co-digestion may pose challenges in terms of thermal energy requirements.

Maintenance of heat exchangers may increase with co-digestion of viscous feedstocks. The heat exchanger surfaces can get coated and be hard to clean.

2.2.2.3 Gas Collection Considerations

A review of existing or planned digester gas collection and transfer piping is crucial to assess whether the gas system can handle the additional gas produced through co-digestion. Unlike natural gas, digester gas is "dirty," containing methane, carbon dioxide, saturated water vapor, hydrogen sulfide and mercaptans, siloxanes, particulates, and many other trace substances. Table 6.2 provides a detailed comparison between natural gas and digester gas to help explain the unique properties of digester gas.

Older digestion facilities may have inadequate gas withdrawal and conveyance piping, especially inside the digester. One POTW reported generating so much gas in their digesters during initial co-digestion operations that their floating digester cover tilted and jammed because the gas withdrawal piping was too small to handle the gas withdrawal, and the gas pressure under the cover became unbalanced (according to Amy Underwood of Downers Grove Sanitary District, 2024).

Conventional digesters operate as very low-pressure gas systems, usually less than 12 inches of water column (w.c.). Thus, digester gas piping must be designed with minimal pressure loss to maximize gas conveyance and avoid operations problems. Pressure loss calculations for digester gas are more

TABLE 6.2 Natural Gas Versus Biogas Properties (Calbry-Muzyka et al., 2022; U.S. EPA, 2020; WEF, 2023b)

Property	Natural Gas	Biogas[1,2]
Methane Content	70%–90%	50 to 75%
Oxygen	0.02%	0-1%
Nitrogen	1.6%	1-3%
CO_2 Content	0.7%	35-40%
H_2S Content[3]	Less than 0.1%	0-1%
Particulates	None	Common
Heat Content	700–1000 BTU/CF	500–700 BTU/CF
Moisture Content	Dry	Saturated[4]
Typical Operating Pressure at a WRRF	2–3 pounds per square inch (psi) to 100 psig	2–3 inches of water column up to 25 psi

Notes.

1. Typical mesophilic anaerobic digestion produced biogas based on typical POTW residential strength influent raw sewage. Biogas production values will vary based on influent feedstock.
2. Conditioning of biogas can change these values. Moisture content can be reduced to near zero as well as particulates.
3. Not mercaptans. Mercaptans may be present in biogas but are added to natural gas for safety.
4. Moisture content, H_2S, and particulates can be reduced to near zero levels via biogas conditioning.

empirical than for air or natural gas. They should have a generous safety factor to account for changes in gas density and pipe roughness over time.

Figure 6.1 outlines a typical digester gas piping system (WEF, 2023a). Generally, digester gas piping must be a minimum of six inches in diameter or larger with gas velocity not exceeding 12 fps. Piping runs should have as few bends as possible and be sloped a minimum of 2% to low points with condensate/sediment traps to facilitate drainage. Gaskets should be a minimum of 1/8-in.-thick Buna-N (Nitrile) (natural rubber and ethylene propylene diene monomer [EPDM] should not be used) and full faced. Outdoor digester gas piping should be heat-traced in northern climates to avoid freezing. Once the digester gas has been treated (filters, dryers, siloxane treatment), the gas behaves more like typical natural gas. A newly developed code that can be referenced for biogas piping design is ANSI/CSA B149.6.

Co-digestion can generate substantial amount of additional biogas but also has the potential for causing foaming in the digester. This can have a detrimental effect on gas collection piping and downstream gas handling equipment. It is recommended to install foam traps/separators on all gas removal piping where it exits the digesters, upstream of any gas handling equipment.

Gas piping material has traditionally been ductile iron pipe (DIP); however, most DIP manufacturers do not support using DIP for digester gas (U.S. Pipe, n.d.). Digester gas typically contains hydrogen sulfide, which, when in contact with water vapor, can create sulfurous acid and corrode DIP pipes and joints over the long term. Welded stainless piping is the preferred corrosion-resistant piping material for untreated digester gas. High-density polyethylene (HDPE) is commonly used for natural gas and is corrosion-resistant

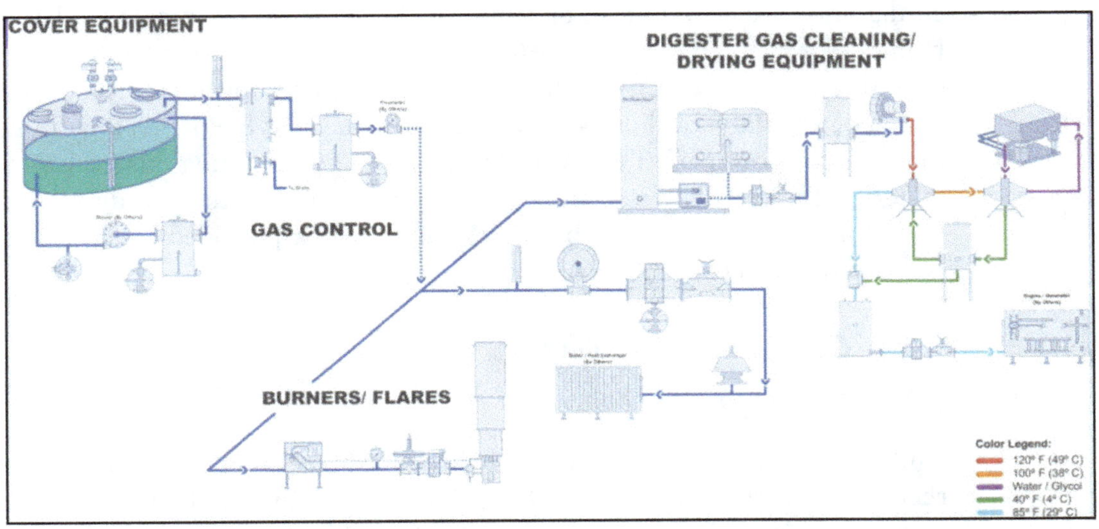

FIGURE 6.1 Typical Digester Gas System (WEF, 2023b)

but not often utilized in facility piping. However, HDPE could be used to transport digester gas underground.

It is important to note that all digester gas safety measures, such as pressure-relief valves, condensate/sediment traps, and flame arrestors, must be provided per National Fire Prevention Association (NFPA) 820, the International Gas Code, and local building codes. Even though additional gas generation and reuse utilization is the goal of a co-digestion program, a functional waste gas burner is still required to flare off any excessive methane gas safely.

An evaluation of gas storage capacity versus gas utilization should also be conducted. Even with consistent feedstock utilization, fluctuations in gas production can hinder end-use as utilization. While gas storage allows for these fluctuations, it's essential to consider its high capital and maintenance costs.

A gas storage analysis should consider the difference in gas volume generated between the estimated maximum gas production rate versus the planned minimum end-use gas utilization rate and whether providing additional gas storage is cost-effective versus flaring. The difficulty in the analysis is estimating a reasonable duration for this condition to occur, but the duration is often economically limited by excessive gas storage costs.

2.2.3 Biosolids Impact

Chapter 4 describes co-digestion's impacts on biosolids in more detail. Biosolids impacts are generally highly variable based on the specific characteristics of each POTW and the feedstocks they co-digest. Generally speaking, co-digestion can be expected to increase biosolids production by 25% to 50% or much higher, depending on the feedstocks being digested.

Biosolids reuse/disposal quality may also be impacted. Particularly, FOG co-digestion seems to impact biosolids quality. One POTW indicated that a high rate of FOG co-digestion (>60% of the total co-digestion loading) has caused its biosolids quality to change; pre-co-digestion, the POTW produced ~70% of its total biosolids as Class A and ~30% Class B. Post-co-digestion, this ratio was reversed. This POTW believes it's not digesting this feedstock material efficiently and is changing its co-digestion operations to digest less FOG material to restore its biosolids quality (data provided by Downers Grove Sanitary District, 2024).

2.3 Co-digestion Handling

2.3.1 Pumping

Sludge pumping can be a complex hydraulic subject that considers sludge rheology or deformation of the sludge fluid. Sludge and feedstock rheology is impacted by pH, temperature, total solids (TS) content, and hydrolysis (digestion) of the sludge (Hong et al., 2017).

Thinner sludge, such as WAS (<2.0% total solids), can typically be assumed to exhibit Newtonian turbulent fluid behavior with reasonably accurate headloss estimations. Thicker sludge, such as primary or thickened sludge (>2.0% total solids), can exhibit non-Newtonian fluid behavior, which requires substantially higher pumping energy to overcome the high shear forces required to start moving the fluid from rest. Viscous feedstocks can also exhibit non-Newtonian hydraulics behavior.

A standard method to address high initial headloss and avoid overloading sludge pump motors is to provide an oversized motor for sludge pumps. In most cases, sludge pumps have low horsepower requirements (typically 5 to 25 hp), and upsizing to the next available motor size is a simple and cost-effective way to avoid overloading the pump motor if the headloss is higher than anticipated.

Positive displacement pumps with large diameter solids passing capabilities are typically preferred for co-digestion pumping. These pumps include plunger (piston) pumps, double-disc pumps, and rotary lobe pumps. These pumps experience more wear than centrifugal pumps, requiring more maintenance, but generally have better solids-handling performance and less downtime due to clogging issues.

In general, centrifugal pumps can be utilized for sludge pumping but should be given careful consideration for solids passage and hydraulics estimations. Large-diameter solids passing centrifugal pumps (such as chopper, screw centrifugal, and vortex pumps) have successfully been used for low TS sludge pumping. However, they should be thoroughly analyzed for high TS sludge (greater than 2.5%) and highly viscous feedstocks.

Sludge grinders may be an upstream consideration for all sludge pumping equipment in order to reduce pump and digester maintenance. However, grinders (and macerators) only reduce debris into smaller pieces, which may cause other problems. Reducing debris getting into the digester is the preferred solution to improving digestion operations.

2.3.2 Feedstock Feeding

Feedstock loading can vary depending on the size of the WRRF and the number of digesters available. Co-digestion can be performed in digesters fed with primary or secondary (WAS) sludge (or both). Whatever pumping method is used, variable-capacity pumping is highly recommended. All the co-digesting POTWs reviewed for this publication reported that the operational concept of slow and steady feedstock pumping into the digesters is the key to providing stable co-digestion operations, even with mixed feedstock material (DGSD, GWA, KWRD, 2024). Depending on the size of the WRRF, it has been observed that feeding the digester slowly but continuously (24/7 operation) results in less process upsets than batch (shift) operation.

For small facilities, it is recommended to initiate co-digestion with the digester that has the highest capacity available. Additionally, utilizing the digester(s) with the best mixing and/or highest amount of gas storage capacity is also recommended. For large facilities with many digesters, which typically may do "round-robin" digester feeding, it may be beneficial to start with one digester, again preferably those digesters with good mixing, gas storage, and work upward to multiple digesters as available.

Feedstock pumping should be introduced as a minimal percentage of the total current digester loading (<10%) while starting co-digestion or when changing feedstock materials. Loading rates can then be ramped up in small increments (10% to 20% intervals) over time (time being a factor of the HRT/SRT of the digester) to increase loading. Digester feed rates can vary between 1.62 to 6.48 kg/m^3/d (~100 to ~400 lbs/1000 cu ft/d) of digester capacity (WEF, 2023a).

When starting co-digestion or changing feedstock loading rates, digester process control variables should be monitored frequently (see Chapter 7 on O&M recommendations).

As will be discussed in the following sections, feedstock storage is critical to allowing consistent dosing of feedstocks into the digesters.

2.3.3 Screening and Grinding

Screening and/or grinding/macerating feedstocks before feedstock storage and co-digestion is highly recommended, especially if it has not been prescreened and/or is stringy and fibrous. Screening is essential to allowing mixing to be efficient as well as reducing clogging problems with pumps, pipes, and valves. Screening can range from a simple bar rack to mechanical screens. The type of screen will be dependent on the size of the facility and the anticipated feedstock material. Feedstock material needs to eventually be reduced to a liquid or slurry for pumping purposes. It should be noted that it is not recommended to dump feedstock material for co-digestion into the POTW's raw sewage preliminary works or primary clarifiers.

Grinding/macerating exposes more of the feedstock's organic matter's surface area, increasing decomposition during digestion. Grinding is typically better at handling and reducing large and hard material, while macerators are better at cutting up soft and stringy material. Another useful item for feedstock receiving is a rock trap, which helps catch hard debris common from receiving FOG from grease haulers. Some manufacturers combine a rock trap with a macerator into one machine.

However, note that effective screening starts with the source of feedstock. It is recommended that strict inspection and testing protocols be established to receive feedstocks that do not require substantial preliminary treatment. While some feedstock may be highly desired for its organic content, spending a lot of POTW resources removing unwanted debris will reduce the efficacy of the co-digestion program.

2.3.4 Piping and Valving

When providing for revised or new digester piping and valves, emphasize installing straight runs and keep bends, tees, and other sharp changes in direction where viscous material can accumulate to a minimum.

Generally, co-digestion piping should be designed to maintain an average velocity of 4–6 fps to provide scouring velocities. Unfortunately, as digester feeding and pumping operations are often batch-processed, settlement of debris and grit in the piping is inevitable. Higher design pipeline velocities will help scour the piping and make cleaning less frequent. Pipe diameters less than 4″ should generally be avoided.

Pipe and fittings should be glass-lined or ceramic-lined ductile iron pipe, Schedule 80 PVC, or stainless steel, depending on the anticipated feedstock material. Special care should be taken to specify corrosion-resistant piping if low or high-pH feedstocks may be handled. Many organic feedstock wastes can have a low pH, which can damage metal components over time. On the other hand, if feedstocks may be abrasive, metal piping is a better choice. Piping gasket selection can be tricky, as it depends on the feed-stock material. For instance, typical FOG is comprised of fatty acids and oils; rubber-based materials are not suitable for fatty acids and EPDM swells when exposed to oils. Fluorelastomers (Telfon) are recommended for all-around chemical resistant material. This material is also recommended for other flexible materials such as power cables.

Provide sufficient flushing connections at the end of pipes, elbows, tees, and other locations to allow flushing, as this will be a critical maintenance item when dealing with viscous feedstocks. Pigging type stations may also be a consideration for longer straight piping runs. It is recommended to work with WRRF operators using clearly labeled piping plans and 3D CAD layout models to define the best locations for flushing/piping connections.

Valves: Valves should be full-port plug or resilient seated gate valves, orientated properly for sludge flow. Butterfly valves, seated gate valves, and other valves with internal recesses should be avoided.

Motor-operated valve actuators are commonly provided to automate co-digestion operations. However, it has been observed that motor operators using limit switches can malfunction as the switches get gummed up from viscous material. Also, as co-digestion piping systems are usually designed to be low pressure (less than 25 psig), valve motor operators sized just for the "system pressure" torque may not have sufficient torque to open or close a valve against heavy, viscous fluid. Similar to pumps, motor actuators should be oversized to provide additional torque. However, torque limiters should also be installed to avoid damaging the valves from the over-torque.

2.3.5 Storage and Equalization

Bulk storage of feedstocks can be critical to smooth co-digestion operations and avoid overloading the digesters. The amount of feedstock storage is highly dependent on the anticipated feedstock loading and the capacity of the digestion system. Especially for new co-digestion and feedstock receiving operations, the amount and type of feedstocks can vary widely. Providing at least 2–4 days of daily storage feedstock volume helps mitigate receiving issues.

Generally, long-term storage of feedstocks is not recommended as decomposition can start to occur, reducing the organic loading to the digester and generating methane or other hazardous gases.

Out-of-service concrete tanks, such as thickeners, equalization tanks, clarifiers, and aerations basins, have commonly been repurposed for feedstock storage. Concrete tanks are typically a good storage medium that can withstand most feedstock. Repurposing can be ideal for keeping capital costs down, especially when a new co-digestion program is initiated. Metal tanks can also be used but may have to be lined against corrosion.

However, based on lessons learned from existing co-digestion facilities, storage tanks should have the following features:

- Provide wall fillets and sloped floors to direct feedstocks to a pump suction and allow washdown and cleaning. Acidic feedstocks may require lining of storage tanks to prevent long-term corrosion.
- Covering the tanks to help mitigate odors. Tanks must also be provided with access ports or hatches for cleaning and vents in case of off-gassing. Feedstock receiving and storage can generate unpleasant odors, especially with FOG material. Passive odor control, such as activated carbon canisters on storage tank vents, are low-cost measures to mitigate odors. More substantial odor control measures may be required depending on the size of the feedstock facility as well as the location of the POTW.
- Provided with liquid-level sensors to measure and track feedstock storage capacity as well as measure digester feed volumes.

- Provided with mixing to keep feed-stock solids in suspension.
- Provide multiple tanks for storage tanks to be out of service for cleaning while maintaining co-digestion operations.
- Electrical: All power and instrumentation cables used for submerged mixers, pumps, valves, and instruments should be extra heavy-duty and corrosion-resistant. One WRRF reported unexpected damage and corrosion on mixer cables that required replacement after only a few years.

  Review the electrical classification requirements per NFPA 820. All anaerobic facilities are rated Class 1, Division 1 or Division 2. Generally, any area within 5 ft of these facilities (or digested sludge processing or digester gas handling) is also classified as Class 1, Division 1, Group D. The area extending between 5 ft and 10 ft from an anaerobic digester is classified as Class 1, Division 2, Group D. Either classification requires all electrical equipment within those areas to be supplied as intrinsically safe, explosion-proof (XP), and/or provided with extensive ventilation. These requirements add considerable capital and O&M costs.

  Most feedstock materials should not be XP classified, but some, such as FOG, may be considered "raw sewage" depending on their source (such as grease traps) and require similar classification.

  Electrical hazard classification can provide challenges regarding locating feedstock receiving, storage, and pumping. Having the feedstock material close to the co-digestion facilities reduces issues with the long-distance pumping of feedstock. However, feedstock facilities situated too close to NPFA 820 classified facilities may require the same classification. These design trade-offs should be carefully reviewed.

Based on the recommended storage tank design criteria, it may be more advantageous and cost-effective in the long term to construct new feedstock storage tanks specifically designed for the purpose. Per Figure 6.2, a specifically designed feedstock receiving and/or storage tank can be designed to receive and screen feedstocks and provide effective mixing.

New digestion programs commonly retrofit or build new storage tanks after several initial years of operation to reduce O&M costs from lessons learned from adapting old facilities. One POTW designed a feedstock loading and storage facility that allowed warm digested sludge to be recirculated around the feedstock, keeping it less viscous and easier to convey.

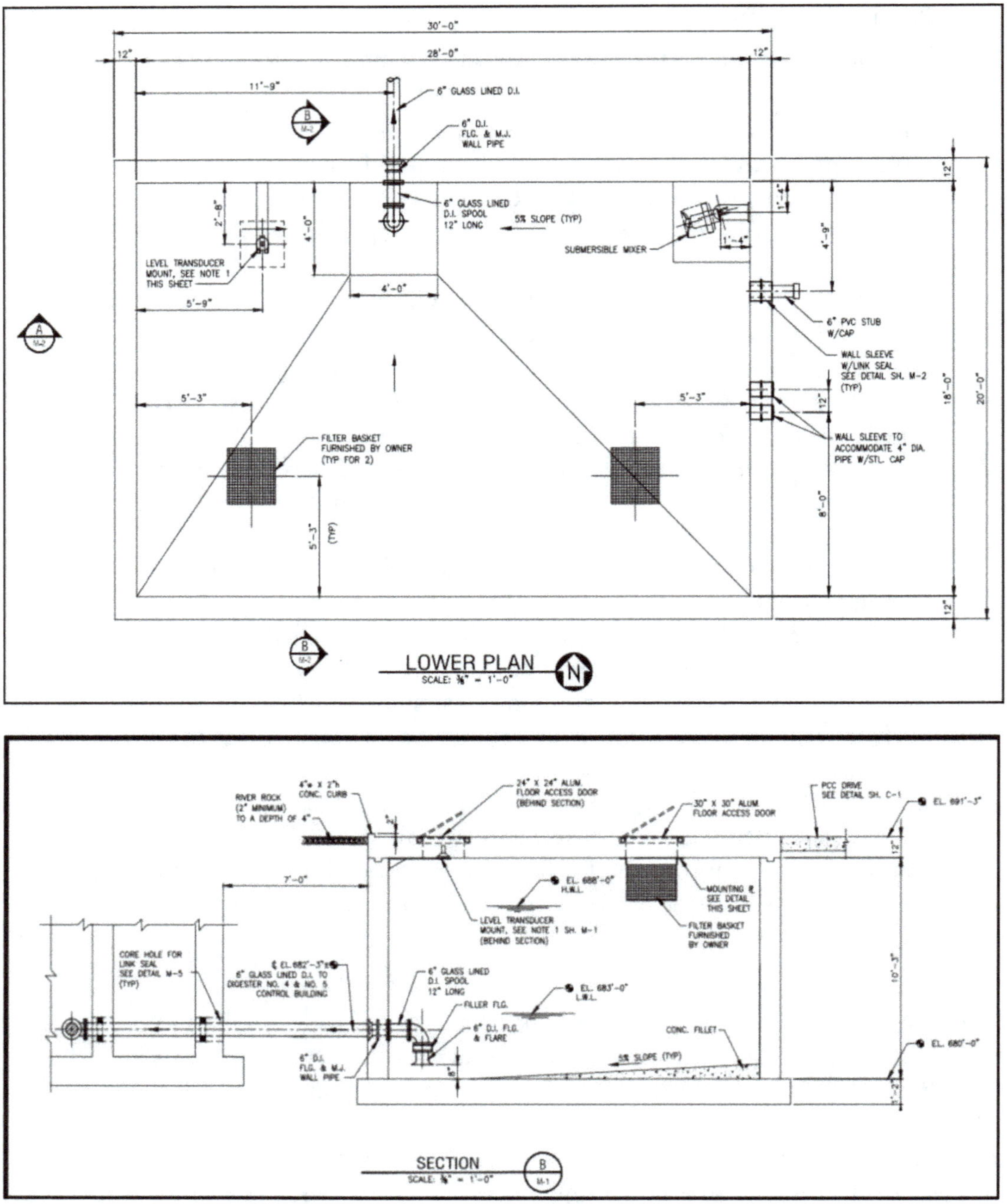

FIGURE 6.2 Receiving/Storage Tank (Reprinted with permission from Downers Grove Sanitary District, 2024)

3.0 REFERENCES

Calbry-Muzyka, A., Madi, H., Rüsch-Pfund, F., Gandiglio, M., & Biollaz, S. (2022, January). Biogas composition from agricultural sources and organic fraction of municipal solid waste. *Renewable Energy*, *181*, 1000–1007.

Great Lakes—Upper Mississippi River Board of State and Provincial Public Health and Environmental Managers. (2014). *Recommended Standards for Wastewater Facilities* (84.32). Health Research, Inc. https://www.health.state.mn.us/communities/environment/water/docs/tenstates/tenstatestan2014.pdf

Hong, E., Yeneneh, A. M., Sen, T. K., Ang, H. M., & Kayaalp, A. (2017, October). The relationship between physico-chemical and rheological characteristics of digested sludge, biosolid, centrate and the effects on dewatering performance (A case study). *Journal of Water Process Engineering*, *19*, 193–204.

Illinois Pollution Control Board. (2024). *Part 370 Illinois recommended standards for sewage works section* (370.830.c.2.A). https://pcb.illinois.gov/documents/dsweb/Get/Document-12042/

Martin, S., & Coker, C. (2016, June). Biogas piping design and safety. *BioCycle*.

National Fire Prevention Association. (2021). Standard for fire protection in wastewater treatment and collection facilities (820).

Schlicht, A. (1999, April). *Digester mixing systems: Can you properly mix with too little power*. Walker Process Equipment.

U.S. Environmental Protection Agency. (1979). *Process design manual for sludge treatment and disposal* (EPA 625/1-79011).

U.S. Environmental Protection Agency. (2020, November). *Anaerobic digester/biogas system operator guidebook*. https://www.epa.gov/agstar/anaerobic-digesterbiogas-system-operator-guidebook

U.S. Pipe. (n.d.). Frequently asked questions. https://www.uspipe.com/technical-tools/frequently-asked-questions-faqs/

Water Environment Federation. (2023a). *Standards for design of water resource recovery facilities* (WEF 8).

Water Environment Federation. (2023b). *Standards for operation of water resource recovery facilities* (WEF 11).

Chapter 7

Co-digestion Systems Operations and Maintenance

Steven Greseth, MBA, PE; Gary W. Neun; Behrang Pakzadeh, PhD, PE; Jamaie Scott, PE; Kyle Stern, PE; & Caroline Wilson, PE

1.0 INTRODUCTION

1.1 What Really Matters: The People

The preceding six chapters have focused on and discussed the biological and chemical aspects of an anaerobic co-digestion system. This chapter is the point where the discussion shifts focus from the "soft-ware" (biomass) to the "hard-ware" (equipment) and the interaction of the utility staff.

The key to this entire process is that the entire utility staff—from the accounting group through the warehouse to the on-site facility labor—should understand the utility objectives and their individual contributions needed to accomplish these tasks. To ensure a common understanding across all groups, the utility management group must present a common:

- Awareness
- Understanding
- Application

The utility mission focused on effective biomass treatment and effective biogas production with incremental efficiency.

1.2 Goal for Operation and Maintenance

The goal for operations and maintenance (O&M) is to enable personnel to monitor and control the natural degradation of the system to maximize the efficiency for the minimum overall costs. These overall costs not only include the day-to-day dollars spent on maintenance, but also the lifetime in equipment (Gerardi, 2003). This is done by:

- Staffing plans
- Development of standard operating procedures (SOP)
- Operational and maintenance training program
- Operational process control
- Process sampling, handling, and analysis
- Process troubleshooting
- Critical review of the treatment equipment

1.3 Operational Changes and Challenges of Implementing Co-digestion

Municipalities with existing anaerobic digestion systems treating municipal biosolids should have existing operational protocols, management systems, and policies in place. These policies must be reviewed when considering implementing a co-digestion process (McKinney, 1962).

1.3.1 Management of Feedstock Stakeholders

The oversight and management of feedstock sources from stakeholders in industry is a time-consuming and management-intense process. Unlike a typical anaerobic digestion process treating just municipal biosolids, many different industry wastes and schedules need to be accommodated and managed.

To effectively consider whether a given feedstock is a good candidate for co-digestion, the following criteria must be considered.

1. Characteristics of the feedstock

   Other sections of this text provide guidelines for effective characteristics of potential co-digestion feedstock. Testing the feedstock will require additional internal or external laboratory resources, particularly if the feedstock has variability in quality. Furthermore, the feedstock may need to be processed or depackaged before its addition to the anaerobic digester. Certain fats, oils, and greases can solidify at lower temperatures and heating of the receiving station may be required. Otherwise, this may lead to blockages in the piping.

   Higher concentrations of nutrients may cause digester process disruptions and/or additional precipitation on process equipment, including struvite and brushite.

Feedstocks may also bring additional inert solids, such as grit in manure or woody material in agricultural waste. These types of inert materials do not degrade in the anaerobic digester and will become the responsibility of the wastewater resource recovery facility (WRRF) to dispose of additional solids. Increased biogas production has the potential to produce more revenue but additional cost of solids disposal should be estimated as well.

2. Transportation logistics

   Introduction of new feedstocks will introduce additional truck traffic to the facility, which can bring safety issues to onsite workers.

   Another key stakeholder for the WRRF is the nearby municipality. Should the feedstock bring offensive odors, neighboring groups may object to this operation. This includes travel routes as new truck traffic may bring odors that are problematic in most areas.

3. Feedstock availability and schedule

   The nature of the industry from which the feedstock comes may mean the feedstock has sporadic availability. Food waste tied to agriculture may only have feedstock available seasonally or the nature of an industry may have out of spec waste in higher quantities due to process issues. The nature of anaerobic digestion requires little change in the overall daily organic loading rates, influent characteristics, temperature, and residence time. These variations in feedstock quantity and quality must be understood to determine co-digestion viability.

1.3.2 Odor Control

Feedstock unloading areas will produce odors. For example, feedstocks which are solid in nature may be unloaded on a tipping floor for inspection before treatment. Liquid feedstocks that are pumped from a truck may have spillage. Water resource recovery facilities, particularly those near the public, should be prepared to utilize, expand, or install existing odor control systems (Gerardi, 2019; Speece, 1996, 2008).

1.3.3 Outreach to Industries

Those in the general industry may not be aware of waste diversion programs or the voluntary co-digestion market. Water resource recovery facilities can charge less than landfills to take on diverted feedstocks and carbon intensity diversion provides positive messaging to the public. However, stakeholder industries must understand the WRRF is not a landfill. Feedstocks have strict requirements that can be rejected by the operations staff for many different reasons, including solids that are too large or inclusion of trash.

2.0 STAFFING PLANS

2.1 Practical Requirements Versus Permit Requirements

The utility staff should clearly understand the facility operating permit(s) criteria and the daily operating plan(s) to adhere to these requirements. Operating permit violations should be the exception, not the rule.

To sustain continuous permit compliance requires attention that, in turn, is focused on the practical requirements to demonstrate compliance. The range of these requirements includes but is not limited to the following:

- Sufficient staff/shift
- Knowledgeable staff
- Staff have both the responsibility and authority to ensure effective biomass stability
- Effective standard operating procedures
- Ready access to tools, consumables, and critical spare parts

2.2 Matching Operation and Maintenance Skills to Design

Matching O&M skills to design is a sensitive point that must be discussed, understood, and resolved by both O&M and design personnel. The designer's role is to develop a facility treatment method, including equipment, which will allow the utility to meet or exceed its current and most extreme future operating permit criteria without considerable modifications.

The utility role is to critically reflect on the utility knowledge base and develop a human resources plan to attain/maintain the quantity of people, with the necessary skills to sustain the utility O&M goals. Utilities need to recognize the workforce pool is becoming more and more shallow, while the individual staffer knowledge requirements are becoming larger and cutting-edge technology focused.

The utility may not be flush with financial resources to continue to perform all work tasks using in-house staff. The utility should consider focusing their resources on the process treatment stream equipment and defer the non-process equipment maintenance (such as high-mast lighting; heating, ventilation, and air conditioning [HVAC]; plumbing; landscaping; and painting, for example) to specialized contractors.

3.0 DEVELOPMENT OF STANDARD OPERATING PROCEDURES

3.1 Definition of Standard Operating Procedures

Standard operating procedures (SOPs) should be the backbone of the actual operations, if maintained and used by the staff. The personnel operating a digester will naturally vary greatly over both the long term (over years) and the short term (during the day). The everyday operator is likely not versed in every part of the digesting system and the facility as a whole and relies on pre-established, step-by-step plans to ensure the system continues to run optimally (Gerardi, 2003). These step-by-step plans, or SOPs, are highly detailed guides to:

- Everyday operations
- Regular maintenance
- Deliveries
- Emergencies

The SOPs should capture almost any situation that will arise. By the existence of these SOPs, there is a plan in place and nothing is left to confusion, potentially leading to poor sampling, unstable conditions, damage to equipment, and so forth. No normal activity should occur that does not have a SOP.

These SOPs are not developed all at once and are not developed in a vacuum. The basics behind the SOPs are established by permitting requirements and regulations and grow until the "final" (final in quotes, because SOPs are constantly updated) version written down by staff. The SOPs are developed by and for all levels of staff (operations, laboratory, maintenance, and management). The staff with hands-on contact to the systems (operations, laboratory, and maintenance) will be the main users of, and thus main authors of, the SOPs.

3.2 Defining, Development, and Utilization of Standard Operating Procedures

Step 1—Purpose

To create SOPs, the first step is to understand what the goal of the SOPs is. This may seem obvious (to successfully operate the digester system) but firm, measurable goals are needed. This includes the final product, laboratory test results, the amount processed, and test results of the health of the digester. The SOPs need to verify that each of these goals will be met in the timely fashion required.

Step 2—Design

In designing and building the digester, plans have been made on how the system will be able to achieve the goals identified in Step 1. Operations and maintenance manuals provide an understanding of the equipment, function, troubleshooting, and so on. The specifications and other design documents provide a systematic understanding of how the system operates (Metcalf & Eddy, Inc./AECOM, 2014). Startup and commissioning demonstrates the functionality of the facility, and ensures goals are met. Operator training teaches how to operate equipment and provides instructions on operations and maintenance. Combined, these are the first versions of the SOPs.

Step 3—Operations

This is where the first “final” version of the SOPs is created. The system is now running in normal conditions. Each type of staff will now create their SOPs for their own positions. As indicated above, these SOPs are not created from scratch, but are adapted, simplified, and organized to be relevant to the staff who will depend on them. This is essential because the personnel present to operate the facility are likely different from the startup and commissioning team. There may be only one or two operators present at a time, where there may have been many more engineers or contractors before. Routines need to be established that will allow the smaller number of personnel to perform all necessary tasks in a way that is efficient and productive. For instance, SOPs need to be written for the once-a-month sampling for some permitting requirements, while other SOPs need to be written for the sampling that occurs daily. Some activities, such as equipment maintenance, will likely be near-mirrors to the procedures established in the O&M manuals. Other activities, such as delivery of chemicals, will likely be established between operators and delivery drivers. These activities are repetitive, and need to be performed uniformly regardless of who performs the task. The SOPs establish a set procedure and make this possible.

Step 4—Development of Standard Operating Procedures

Documentation of SOPs is highly unique for each location. As discussed above, the main objective is to make these documents useful. This can mean:

- Each step has a picture included with markups indicating the location of various valves, sample locations, etc.
- Text only step by step
- Links to saved videos

Whatever format is agreed upon, the SOPs should be a step-by-step guide that walks through each task. To ensure these are complete and accurate, ideally these should be written while performing the task. For example, the SOP for a pump oil change is likely based on the Equipment O&M manual instructions, but it should be written in the field while performing the oil change. This not only ensures no steps are missed, but also this is a final quality check that guarantees each SOP is correct for the system. Since SOPs are written in the field, the supervisor needs to review the SOPs as they are completed. Standard operating procedures are living documents and are the guide that enables successful operation of the digester.

Step 5—Accessibility and Utilization

It cannot be overstated that the best plan in the world is useless if it is not used. There are hundreds of SOPs, and some are only used on rare occasions. For these to be utilized, the plans need to be stored in a manner that is functional. Software is available that can be used to grant a group access to these changing documents. Consistent formatting and naming schemes also aid in ease of use. As personnel change (employees quit, are reassigned), it is essential that SOPs are not lost with the employee. The format of this access is less important than all staff buying into and utilizing the plans to the fullest.

3.3 Updating Standard Operating Procedures

3.3.1 When to Update Standard Operating Procedures

As essential as creating SOPs is updating SOPs. If SOPs are not updated constantly as conditions change, they quickly become obsolete and are thus not used. If the SOPs are being utilized by staff regularly, updating the SOPs as changes occur should be relatively natural. To ensure this occurs, supervisors should review the SOPs on a regular basis or when significant changes occur. Updating the SOP is not a negative action, and normally does not indicate mistakes with the existing SOP. Rather, updating the SOP is a natural reflection of a change in (included but not limited to):

- Permit requirements
- Personnel
- Equipment
- Scientific discoveries
- New construction
- Vendor/supplier
- Process fluid

Depending on the impetus of the change the type of staff, the amount of discussion that must be had before changing the SOP varies. Table 7.1

TABLE 7.1 Standard Operating Procedures Update Responsibilities

Issue	Examples	Responsible	Inform
Permit Requirements Change	• Federal regulations • State regulations • Local regulations	Management	Operations Laboratory
Personnel Change	• New personnel hired • Personnel quit • New shift structure	Operations Laboratory Maintenance	Management Operations Laboratory Maintenance
Equipment Change	• New equipment • Damaged equipment • New technology	Operations Laboratory Maintenance	Management Operations Laboratory Maintenance
Scientific Discovery	• New tests • New testing equipment	Laboratory	Management Operations
New Construction	• New equipment • Changing facility conditions • Changing process fluid • Change in operations	Management Operations Laboratory Maintenance	Management Operations Laboratory Maintenance
Vendor Changes	• Switch to new supplier • New delivery procedure • Change in product provided	Operations Laboratory Maintenance	Management
Process Fluid Changes	• Flow from new location • Change in existing flow characteristics	Operations Laboratory	Management Operations Laboratory

shows some potential reasons to update SOPs, examples, the type of staff responsible for the updates, and the type of staff that need to be informed that the updates occurred.

3.3.2 How to Update Standard Operating Procedures

When updating the SOPs, the process changes depending on the significance of the update. Changes need to be indicated by a revision date on each page (e.g., "current as of June 2024"). This allows a reviewer and the user to know if the changes are current, or if a new update may be needed. Updating does not need to be a formal or complicated process; in fact, an update may not even include changes. The SOP simply needs to be reviewed and verified to be accurate on a regular basis.

New technology, for example a new piece of laboratory testing equipment, requires the approval of management as well as the knowledge of the laboratory staff to utilize the equipment successfully. If the new equipment makes in-house testing possible, it may also affect the operations staff in sampling procedures. A smaller change, for example a new type of delivery truck, would only require coordination and updating from the operations staff. New construction would require changes from all staff, likely continuing to change over the duration of the construction. In summary, updating SOPs is a constant process involving all levels of staff. The key is constant communication.

The operation of a digester is fluid and changing, and thus the SOPs must also be fluid and changing. When the SOPs are developed with care and are updated as situations change, the result is an extremely valuable set of documents that enable the system to be run by anyone qualified. This helps prevent the sticking point of one person being essential to the smooth operations of a system, which is an unreasonable expectation of any person and an unreasonable reliance on behalf of the facility. The SOPs are updated as issues arise, and thus prevent items from slipping through the cracks (for example, a permit requirement due at the end of the year). By updating the SOPs constantly, the SOP remains a useful document that was developed for and applies directly to the individual system.

4.0 OPERATIONS AND MAINTENANCE TRAINING PROGRAM

4.1 Matching Training to Design

Training is the act of investing resources into staff to develop skills and perspectives needed to operate and maintain the facility safely and effectively. Effective training programs equip staff with the knowledge necessary to operate, maintain, and troubleshoot components and processes within the facility.

The role of the designer is to develop a facility treatment method and determine equipment that will allow the utility to meet or exceed its current and most future operating permit criteria without significant modifications. The designer should specify the inputs and outputs of the design, such as the feedstock specifications.

The role of the utility is to critically reflect on the utility knowledge base and develop a human resources plan to both attain and maintain the quantity with the necessary skills to sustain the utility O&M goals. Utilities must ensure that there is enough overall staff and knowledgeable staff available to carry out the facility functions. The workforce pool is becoming more and more shallow while the staffer knowledge requirements are becoming larger and more cutting-edge, technology focused.

The first step developing an effective training program is understanding the roles of facility staff. Table 7.2, below, shows the typical responsibilities of staff. In developing a training program, the important information relating to an employee's specific job function should be identified. There is no reason to teach an operator how to change a programmable logic controller card or how to carry out laboratory testing. However, all staff need to be knowledgeable of the "big picture." When training is insufficient, preventable mistakes may occur. For example, laboratory staff should be taking samples around the facility, and if operators instead are taking samples, they may leave bottles out for a few hours before sending the bottles to the lab for dissolved oxygen and pH testing. The chemistry inside the bottle will change in a matter of minutes and that is something a laboratory technician should recognize while an operator may not.

This can, in part, be covered by the development of SOPs which is covered in Section 3.0, but mitigation of future issues can also be achieved with a holistic understanding of the system and its goals. Overall goals of the facility can include achieving the greatest volumetric biogas production, highest methane concentration by volume, and the mitigation of odors increasing favorable public perception. The utility management group should present a common:

- Awareness
- Understanding
- Application

Utility mission should be focused on meeting the facility goals with incremental efficiency.

The utility may not have the financial resources to continue to perform all work tasks using in house staff. The utility should consider focusing their resources on the process treatment equipment and defer to specialized contractors non-process equipment maintenance, such as:

- High mast lighting
- HVAC
- Plumbing
- Landscaping
- Painting

TABLE 7.2 Typical Responsibilities of Staff

	Operations	Laboratory	Maintenance	Management
Typical Responsibilities	• Process control • Ensure healthy biomass • Cleaning and calibration of instrumentation • Rejection of out of spec feedstocks	• Take samples • Test important constituents such as pH, alkalinity, VAs, etc. • Communicate trends in produced data to operations	• Lubrication of moving parts and pumps • Replacement and repair of equipment • Cleaning and calibration of instrumentation	• Ensure the facility is meeting its overall goals • Create training program • Ensure the facility personnel have the tools and skills needed to carry out work • Organize feedstock agreements
Training Topics	• Understanding process indicators • Understanding trends in data and correction strategies • Operations Safety	• Standard Methods and laboratory operating procedures • Laboratory Safety	• Creating and following maintenance schedules • Site Safety	• How to create a training program • Training topics • Staff communication

For more complex equipment, it is recommended that a specialist contractor perform the work. Saving money and time is unlikely when training the maintenance team on such complex equipment.

4.2 Targeted Hands-On versus Classroom

The personnel are the lifeblood of the facility. Finding the most effective training method will pay dividends.

Training earlier is better. Providing staff training earlier helps mitigate undesirable habits that can form without proper training.

Training can take on many different forms. People learn differently. Some learn visually, by hearing, by reading, or by doing the work. A training program should include all of these training techniques such that different staff with different learning types can effectively apply information.

Make sure the training is applied. When giving trainings through presentations and readings, you must ensure the guy in the chair learns the material by applying the information. Rather than thinking of the classroom and hands-on work as competing training styles, both should be used to provide a well-rounded training program.

The best time for staff to learn about equipment is during the startup of the facility. The startup staff will be the most knowledgeable on the system and process. The utility staff should shadow and learn as much as they can. It is imperative that the utilities are active in engaging the designers in training staff. The utilities need to understand what resources are available from designers and their own people. The utility and designer can then create a training program to patch the holes. The utility needs to push the designer on what utility staff should be aware of, who will give that level of understanding, and who is going to monitor them to ensure the work is getting done correctly.

4.3 Match CEUs to Training Program

Continuing education units (CEUs) are required by some states to advance operator licenses. Continuing education units provide motivation both in operators seeking certification and in utilities providing resources for maintaining license requirements.

Typically, it is not difficult getting a training program approved for CEUs. The training materials can be sent to the relevant state agency for approval. This group will review and see if the training follows their guidelines. If the facility is in a state that requires CEUs for operator certification, it is often worthwhile for the utility to design and get a certification process approved. This can mitigate travel costs and more facility-specific topics can be covered. This will give third party validation of the training materials and can help the utility when it comes to future permitting.

While maintaining CEUs, it is important to ensure that after the training, the operations staff have learned and taken value from the course. Operations staff are rarely checked to ensure they have learned the material, which can leave holes in understanding.

5.0 OPERATIONS PROCESS CONTROL

Operation and maintenance of an anaerobic co-digestion facility is similar to that of an anaerobic digestion facility that solely treats primary and/or waste-activated sludge. The only difference between the conceptual design of these facilities is the biological/chemical aspects of the organic loading. For a co-digestion facility, the O&M staff must recognize that the system may be treating more than two organic loading streams, the individual constituents of which may increase or decrease on a day-to-day basis. This variability poses a unique process control challenge for co-digestion facilities.

The co-anaerobic digestion facility operations limit are originally determined by the designer. The designer determines the "normal" daily organic loading rate as well as the "maximum day" loading rate to determine the size and quantity of the treatment pumps, piping, mixers, heaters, and tanks. The daily organic loading rate is normally presented in terms of pounds of volatile solids (VS) per thousand cubic feet and/or pounds of chemical oxygen demand (COD) per thousand cubic feet.

The facility O&M staff need to recognize the anaerobic digester designer provided specific design/operations key performance indictors (KPIs) all within the operations control abilities of the O&M staff as follows:

- Organic loading rate per 1000 cu ft
- Mixing energy
- Temperature—maintain the temp within the designer specified range

5.1 Organic Loading Control

The most important job of an operator at a co-digestion facility is to maintain the "normal" daily organic loading rate within ±10% of the operations limit. The organic loading rate is the most critical process control factor that the operator can manipulate (Gerardi, 2003; Water Pollution Control Federation, 1987).

The process of anaerobic co-digestion involves the biological degradation of complex organic solids and the reduction of pathogenic organisms within an anaerobic environment. Anaerobic digestion is dependent upon the activity of the microbe population inside the digester. Operations process control of the digestion system requires operator and laboratory technicians to perform basic laboratory analysis to monitor biomass activity and meet state/federal reporting requirements.

Testing of co-digestion feedstocks is imperative to understand the increase in loading rate per unit feedstock. Proper field sampling, field data collection, laboratory staff analytical technique, and data trend charting enhance anaerobic co-digester performance and operation.

5.2 Mixing Control

There is controversy regarding the input mixing energy to an anaerobic digester tank and mixing control operation. Generally, mixing of the digester occurs constantly. Proper mixing allows contact of the organic material with the biomass. If the digester is not operated in this manner, most of the time, the driving factor for the discussion was "excessive" foaming of the anaerobic digester and the supposed reason was "excessive" mixing.

"Excessive" foaming events are corrected by returning the tankage to the specific designer mixing criteria, concurrent with organic loading control (McKinney, 1962). Losing the ability to heat the tankage or "feed" the biomass simply slows down the biological reaction rates and biogas production. Returning the biomass to the correct temperature range or organic loading rates is readily resolved over time—one-degree temperature increase per 24 hours. Returning to the normal daily organic loading rate may take as long as 45 days. Mixing also assists in maintaining the overall anaerobic digester tankage temperature. Biological commissioning of anaerobic digesters can be completed by simply mixing at the designer specified rate plus organic loading control and no external heating, should there be heating system equipment startup issues.

Mixing may also be done in a reverse mixing mode. This can be done weekly or monthly, as given by the designer, and can minimize grit and inert solids that build up in the digester.

Losing mixing energy is an operations disaster that impacts both temperature control and organic loading. Restarting mixing is best done in stages; assuming there is more than one anaerobic digester mixing system equipment, then start units in series, one every 8 hours, until all designer specified units are on-line.

6.0 PROCESS SAMPLING, HANDLING, AND ANALYSIS

This section discusses data collection, sampling points, sampling basics/frequency, lab procedures, and data interpretation.

6.1 Sample Collection and Handling

Data should be collected, analyzed, and inputted to the appropriate trend chart for each organic loading source and individual anaerobic co-digester. The data points should include information on: volatile acid, alkalinity, volatile acids/alkalinity ratio, pH, temperature, total solids, VS, and ammonia, if required.

The following is a list of data points and the expected operations ranges for each parameter. If data collected from samples falls outside these ranges, re-confirmation analysis should be performed to verify test results.

Note: "normal" parameter data ranges for anaerobic co-digestion facilities are defined as values that produce the greatest quantity and best quality of biogas.

- Volatile acids (VA)—The VA concentration as acetic acid ranges from 50 to 330 mg/L. Depending upon the types of organic loading sources the lower or upper limits for "normal" VA concentration may be significantly lower or higher.
- Alkalinity (ALK)—The total ALK concentration as $CaCO_3$ ranges from 1500 to 5000 mg/L. Depending upon the types of organic loading sources the lower or upper limits for "normal" ALK concentration may be significantly lower or higher.
- Volatile acids/alkalinity ratio (VA/ALK)—The VA/ALK ratio for a well-operated system is generally considered to be consistently less than 0.1. Depending upon the types of organic loading sources, the lower or upper limits for "normal" VA/ALK ratio may be significantly lower or greater.
- pH—The pH ranges from 6.5 to 7.5.
- Temperature—The temperature ranges from 92 to 98 °F for mesophilic digesters. Temperature ranges from 122 to 140 °F for thermophilic digesters.

Each facility must determine the required process control range. Figure 7.1 shows a diagram with reasonable sample values and ranges.

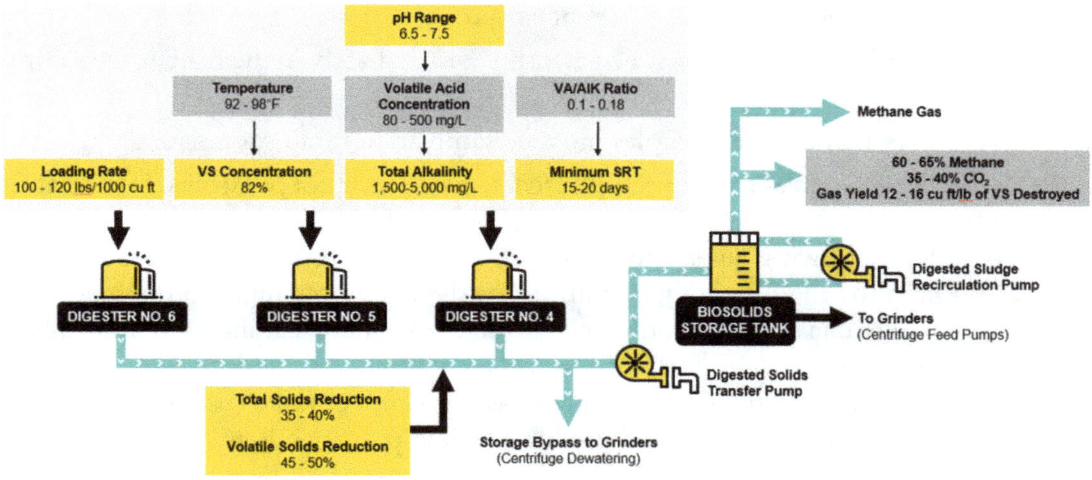

FIGURE 7.1 Digested Solids Process Flow

The following are feedstock-dependent parameters that must be monitored as a part of the operations process. Normal ranges will vary for each facility.

- Total solids concentration of the various organic loading sources (or biosolids) input and system output.
- Volatile solids concentration of the various organic loading sources (or biosolids) input and system output.
- Chemical oxygen demand concentration of the organic loading input and system output.

Run charts of the volatile suspended solids, total daily feed sludge flow and VA/ALK ratio illustrate both positive and negative trends.

6.2 Sampling Analysis

Sludge sampling is vitally important to monitoring the operations health of the co-digestion system. Each digester should have one influent and one discharge sampling point. The mixed digested sludge should also be sampled.

Proper sampling is critical for effective anaerobic digester process control monitoring. It is critical that the operating staff understand these basics for effective sampling:

- All sample containers must be cleaned thoroughly before and after use.

- The sample must be representative.
- pH analysis must be performed immediately in the field at the sample point.
- All digester samples must be sampled in rapid succession.
- Samples must be refrigerated if analyses are not performed immediately.

The following subsections of process control parameters are utilized to monitor the operating conditions of the anaerobic digestion process. The rate of change of an individual parameter is more significant than its absolute value.

Process control trend charting is the best method to monitor anaerobic digester performance and predict possible process upsets.

6.2.1 Digester Influent

This sample should be collected from the influent organic stream.

Temperature	Frequency = Once/shift
pH	Frequency = Daily
Volatile acids (VA)	Frequency = no less than 3/week
Alkalinity (ALK)	Frequency = no less than 3/week
VA/ALK ratio, calculated	Frequency = no less than 3/week

6.2.2 Mixed Digested Sludge

This sample should be collected from the sludge recirculation line before the digested sludge is mixed with the influent organic loading flow.

Temperature	Frequency = Once/shift
pH	Frequency = Daily
Volatile acids (VA)	Frequency = no less than 3/week
Alkalinity (ALK)	Frequency = no less than 3/week
VA/ALK ratio, calculated	Frequency = no less than 3/week

6.2.3 Digested Sludge

This digested sludge should be collected from the digested sludge holding tank or the sludge overflow box.

Total solids (TS)	Frequency = no less than 3/week
Volatile solids (VS)	Frequency = no less than 3/week
Flow (gallons)	Frequency = no less than Daily
pH	Frequency = Daily

6.3 Sample Data Trend Charting and Data Interpretation

Data is collected at the frequency indicated previously. Again, the rate of change of the collected data is more important than the absolute value of the data collected. The operator compares the most recent data collected with previous data to verify the present data point is within a reasonable range of previous data. If the data is not within an acceptable range, another sample is taken and analyzed to confirm results.

If data is satisfactory, it is trended with previous data to gauge overall anaerobic digester performance. Corrective actions must be initiated if performance is unsatisfactory.

The real-time performance of the anaerobic digester should be monitored using the following field process control parameters:

- Carbon dioxide is an easily measured component of the digester gas. Because the sum of the CO_2 and CH_4 is approximately 100%, the amount of CH_4 can be estimated by measuring the CO_2. As the CO_2 concentration increases above 30%, the condition of the methane-forming microorganisms is declining and the digester may be trending toward excessive acid production.
- Observation of the waste gas burner flame is an unconventional process monitoring parameter. A yellow flame with blue at the base is "normal." When the flame will not stay lit and "too much" blue is present, the CO_2 concentration is increasing. An orange flame with smoke indicates a high sulfur content in the digester gas.

Evaluating the process control data is the second critical function for the operating staff. The use of trend charts will visually assist the operator to determine the tendencies of the anaerobic digestion system and allow for proper operations adjustment.

The operator should concentrate on controlling three aspects of the anaerobic digester system:

- Hydraulic detention time (HDT)
- Digester organic loading rate
- Volatile acids/alkalinity ratio

6.3.1 Hydraulic Detention Time

The HDT is used to monitor the detention time of the digester volume displaced by the daily wasting volume. Detention time is critical to meeting the U.S. EPA 503 regulations related to anaerobic digestion.

$$\text{HDT, days} = \frac{\text{Anaerobic digester volume (gals)}}{\text{Combined daily waste sludge flow rate (gal/day)}}$$

A typical target HDT range of 15 days is a common operations goal for anaerobic co-digestion facilities.

The following example calculation can be used to guide an operator in the calculation of their required wasting rate to maintain their target HDT of 15 days.

Known: Anaerobic digester volume of 293 964 gallons
Find: Combined daily waste sludge flow rate (gal/day)
Where:

$$15 \text{ days} = \frac{293\ 964 \text{ gallons}}{\text{Combined daily waste sludge flow rate (gal/day)}}$$

$$\text{Combined daily waste sludge flow rate (gal/day)} = \frac{293\ 964 \text{ gallons}}{15 \text{ days}}$$

Combined daily waste sludge flow rate (gal/day) = 19 598 gallons/day/ anaerobic digester at 15 days HDT.

To increase the quantity of sludge wasted without disrupting the anaerobic process or exceeding the HDT, the operator should increase the either the primary clarifier solids or waste activated sludge solids concentration. The benefits to solids thickening include:

1. A reduction in sludge volume directly reduces the amount of heat required to raise the incoming sludge to the digester operating temperature.
2. An increase in feed solids concentration decreases the volume of solids entering the digester, which increases both the hydraulic retention time (HRT) and sludge retention time (SRT), which promotes better digestion and a higher volatile solids reduction.

Loading the digester with the highest concentration of combined solids maintains the proper HDT, reduces energy requirements, minimizes internal digester temperature fluctuations, and stimulates positive anaerobic microorganism activity.

6.3.2 Sludge Thickening

Many waste facilities have the capacity to discharge primary sludge or waste activated sludge (WAS). The primary sludge and/or WAS concentration is normally less than 1.0% solids or 10 000 mg/L. The thickened primary sludge/WAS sludge discharging to the anaerobic digester should be approximately 2% to 3% solids.

The operator's wasting concerns should center on removing a specific quantity of WAS per day to maintain a specific operating SRT in the secondary biological treatment train. The anaerobic digester requires a specific range of volatile solids and HDT to properly process the sludge.

The following example calculations can be used to determine the thickened sludge feed rate to an anaerobic digester based on a known WAS wasting rate.

$$\text{lbs/day Solids in WAS} = \text{lbs/day solids in thickened sludge}$$

Known: WAS flow = 157 000 gal/day
Percent of WAS Total Suspended Solids = 0.6% or 6000 mg/L
Thickened sludge concentration = 3.0% or 30 000 mg/L
Where:

$$(\text{WAS gal/day}) \times (8.34\ \text{lbs/gal}) \times \frac{(\%\ \text{solids})}{100} = (\text{X gal/day}) \times (8.34\ \text{lbs/gal}) \times \frac{(\%\ \text{solids})}{100}$$

$$(157\ 000\ \text{gpd}) \times (8.34\ \text{lbs/gal}) \times \frac{0.6}{100} = (\text{X gal/day}) \times (8.34\ \text{lbs/gal}) \times \frac{3.0}{100}$$

$$(157\ 000\ \text{gpd}) \times (8.34\ \text{lbs/gal}) \times 0.006 = (\text{X gal/day}) \times (8.34\ \text{lbs/gal}) \times 0.03$$

$$\frac{(157\ 000\ \text{gpd}) \times (8.34\ \text{lbs/gal}) \times 0.006}{(8.34\ \text{lbs/gal}) \times 0.03} = (\text{X gal/day})$$

31 400 = (X gal/day) or 31 400 gallons of thickened sludge discharging to the anaerobic digester

6.3.3 Digester Loading Rate

One effective means to control the anaerobic biological digestion process is by adjusting the VS loading rate. Uniformity and consistency are necessary operations concerns for effective anaerobic digester operation. Changes in feed solids volume, concentration, temperature, or withdrawal rates can negatively impact the digestion process and thus must be done gradually. It should also be noted that for facilities with more than one anaerobic digester inflow, consideration should be given to having the flows tied into one small-volume, well-mixed, organic loading tank.

Both primary and WAS sludge require blending with active recirculated digester sludge reducing the risk of short circuiting due to thermal settling of the colder sludge. Common cause of organic overload includes excessive loading due to infrequent feedings. It is important that the sludge recirculation pumping system should remain at a constant and continuous speed.

The following example calculations can be used to support an operator in the calculation of digester loading rate:

$$\text{Digester Loading, lbs VS/day/cu ft} = \frac{\text{VS added, lbs day}}{\text{Digester Volume, cu ft}}$$

$$\text{Digester Loading, lbs VS/day/cu ft} = \frac{(\text{Combined Sludge, lbs/day}) \times \frac{(\%\text{Solids})}{100} \times \frac{(\%\ \text{VS})}{100}}{\text{Digester Volume, cu ft}}$$

Known: 31 000 gpd combined sludge feed rate
4% solids
65% Volatile Solids
107 000 cu ft digester volume

$$\text{Digester Loading} = \frac{(31\ 000\ \text{gpd}) \times \frac{(4)}{100} \times \frac{(65)}{100}}{107\ 000\ \text{cu ft}}$$

$$\text{Digester Loading} = \frac{6722}{107\ 000}$$ or a digester loading rate of 0.06 lb VS/day/cu ft

This calculated loading rate should be compared to the design loading rate as specified by the designer. Operators should not vary the daily loading rate by more than ±10% from the design basis.

6.3.4 *Volatile Acids/Alkalinity Ratio*

The volatile acid/alkalinity ratio is an indicator of the progress of digestion and the balance between the acid fermentation and methane fermentation microorganisms. Careful monitoring of the rate of change can indicate a problem before a critical change in pH occurs.

$$\text{Volatile Acids/Alkalinity ratio} = \frac{\text{Volatile acids, mg/L}}{\text{Alkalinity, mg/L}}$$

Increase to or above ratios of 0.3 to 0.4 indicate an anaerobic digestion process upset and the need for corrective action. These corrective actions should include decreasing sludge withdrawal rate, extending mixing time and maintaining the internal digester sludge temperature at 95 °F ± 1 °F.

Increases to or above 0.5 ratio indicate an anaerobic digestion process upset and the need for corrective action. These corrective actions should include requiring the addition of alkalinity and decreasing the sludge loading rate. In the event that alkalinity addition is required, volatile acid concentration must be used to calculate the proper concentration.

Increases to or above 0.8 ratio indicate an anaerobic digestion process upset with pH depression and inhibition of methane production occurring. The anaerobic process requires the addition of alkalinity and the reduction of VS loading to 0.01 lb VS/cu ft/day until the VA/ALK ratio drops to 0.5 or below. Volatile acid concentration must be used to calculate the proper concentration prior to alkalinity addition.

In monitoring digester temperature, operators should note that if temperature adjustment is deemed a necessary corrective action, the anaerobic digester temperature should not vary more than 1 °F over a 24-hour period.

The following example calculation can be used to support an operator in determining the required alkalinity addition during a process upset to neutralize volatile acids.

6.3.5 Alkali Dosage Required for Neutralization

Lime can be added to a digester to neutralize volatile acids on a 1:1 basis. It should be noted that lime is less commonly used than other alkali like sodium bicarbonate because of the issues associated with overdosing. Nevertheless, it is helpful for illustrative purposes given that the volatile acid to lime dosage rate is 1:1.

$$\text{Volatile acids, lbs} = \text{Lime required, lbs}$$

$$(\text{Volatile acids, mg/L}) \times (\text{Digester volume, MG}) \times (8.34\ \text{lbs/gal}) = \text{lbs of volatile acid in the digester} = \text{lbs of lime required}$$

Known: VA = 450 mg/L

$$\text{Digester volume} - 0.293\ \text{MG}$$

$$(450\ \text{mg/L}) \times (0.293\ \text{MG}) \times (8.34\ \text{lbs/gal}) = \text{lbs of volatile acid in the digester} = \text{lbs of lime required}$$

$$1099\ \text{lbs of VA} = \text{lbs of lime required}$$

This tells us that 1099 lbs of volatile acid can be neutralized by 1099 lbs of lime.

To utilize other sources of alkali for anaerobic digester dosing, use the following formulas to convert the alkali to a volatile acid concentration as calcium carbonate:

$$\text{Anhydrous ammonia (lbs/day)}^{*} = (\text{VA, lbs/day}) \times 0.34$$

$$\text{30\% ammonia solution (30\% ammonium hydroxide) (gal/day)} = (\text{VA, lbs/day}) \times 0.28$$

$$\text{Anhydrous soda ash (100\% sodium carbonate) (lbs/day)} = (\text{VA, lbs/day}) \times 1.06$$

$$\text{50\% sodium hydroxide solution (lbs/day)}^{**} = (\text{VA, lbs/day}) \times 1.60$$

$$\text{Quicklime (100\% calcium oxide) (lbs/day)} = (\text{VA, lbs/day}) \times 0.56$$

$$\text{Hydrated lime (lbs/day)} = (\text{VA, lbs/day}) \times 0.74$$

$$\text{Sodium bicarbonate (lbs/day)} = (\text{VA, lbs/day}) \times 1.68$$

*It should be noted that the use of anhydrous ammonia as an alkali should be carefully considered with respect to facility discharge permit limits for ammonia.

**Typically, 50% sodium hydroxide solution is purchased in totes or drums, so it may be more helpful to the user to consider the dosage rate in terms of volume rather than weight. The following conversion can be used to calculate gallons of 50% sodium hydroxide required based on the volatile acid concentration:

$$\text{50\% sodium hydroxide solution (gal/day)}^{**} = (\text{VA, lbs/day}) \times 0.125$$

6.3.6 Percent Volatile Solids Reduction

Volatile solids reduction is an indicator of proper anaerobic digestion and a requirement of the U.S. EPA 503 regulations. The U.S. EPA 503 regulations minimum VS reduction requirement is 38% to meet vector attraction reduction requirements.

To calculate VS reductions, the operator should collect a sample of the *feed sludge* influent and the *digested sludge* effluent for total solids and VS analysis. The results of the laboratory analysis are used in the following formula. The size and complexity of the equation belies the actual ease of calculation.

An operator can use the following example calculation to provide guidance on the calculation of percent VS reduction at their facility:

$$\% \text{ VS reduction} = \frac{(100) \times \dfrac{(\%\text{VS in}) - (\%\text{VS out})}{(\%\text{VS in}) - [(\% \text{ VS in}) \times (\%\text{VS out})]}}{100}$$

Known: % VS in = 70
% VS out = 52

$$\%\text{VS reduction} = \frac{(100) \times \dfrac{(70) - (52)}{(70) - [(70) \times (52]}}{100}$$

$$\%\text{VS reduction} = (100) \times \frac{(70) - (52)}{(70) - 36.4}$$

$$\%\text{VS reduction} = (100) \times \frac{18}{33.6}$$

$$\%\text{VS reduction} = (100) \times 0.535$$

$$\%\text{VS reduction} = 53.5\%$$

In this example, the facility would have met the minimum VS reduction standards as outlined in U.S. EPA 503.

7.0 PROCESS TROUBLESHOOTING

7.1 Process Control Trouble Shooting

Although operators of co-digestion systems are advised to operate their systems using the operating parameters as outlined in Section 6.0, they may find themselves in a situation in which their KPIs begin to stray from the best practices as outlined in this manual. This is especially true for co-digestion facilities, which may experience fluctuations in feedstock that can influence overall performance when compared to traditional anaerobic digestion facilities.

The key to successful operation of a co-digestion facility is to recognize and appropriately respond when digester performance begins to stray from target operating conditions. The following section is intended to help operators identify when they have a problem with their digesters and adjust operating parameters as required to rectify the situation.

In the context of anaerobic digestion processes (Table 7.3), operators often encounter challenges that impact system performance. Excessive biogas pressure can result from organic overload, and the solution involves activating the waste gas flare to release excess pressure. When the waste gas flare fails, potential causes include lack of power, pilot ignition failure, or cycling issues. Remedies include stopping organic loading, checking fuses, resetting motor contactors, inspecting spark plugs, verifying gas flow, examining solenoid valves, and adjusting pilot gas supply.

The pH and alkalinity levels are critical for process stability. If pH and/or alkalinity are too low, it may indicate organic overload or incorrect chemical feed. In such cases, adding caustic accordingly can help restore balance. Conversely, high pH and alkalinity may result from excessive chemical feed. Operators should ensure that caustic dosing pumps are turned off on the human-machine interface and in the field.

Effluent quality parameters from the treatment facility and co-digestion substrate also require attention. When volatile solids in the influent feedstock stream exceed the target, organic overload is likely. Lowering the feedstock flow rate can mitigate this issue. Similarly, if the feed COD surpasses the target, reducing the feed flow is recommended. Lastly, maintaining the VA:ALK ratio within the desired range involves adjusting the feedstock flow based on organic load.

TABLE 7.3 Anaerobic Digestion Process Control Troubleshooting

Condition	Possible Causes	Remedial Measures
Excessive Biogas Pressure	• Organic overload	• Activate waste gas flare
pH and/or Alkalinity Out of Range (low)	• Organic overload • Incorrect chemical feed	• Add alkali
pH and/or Alkalinity Out of Range (high)	• Excessive chemical feed	• Stop chemical feed
Feed Volatile Solids Greater Than Target	• Organic overload	• Lower feed flow
Feed COD Greater Than Target	• Organic overload	• Lower feed flow
VA:ALK Ratio Out of Range	• Organic overload	• Adjust feed flow

Remember that these troubleshooting steps should be tailored to the specific system and conditions to ensure efficient anaerobic digestion processes.

Thermal hydrolysis process (THP) may be implemented upstream of a co-digestion system to improve various aspects of the digestion process including biogas production, sludge dewaterability, and allowable loading rate. For facilities that utilize THP as pretreatment, Table 7.4 can be used for troubleshooting.

7.2 Digested Sludge Troubleshooting

The condition of the sludge fed to and circulated within an anaerobic digester strongly influences the health of the operating system. The quality of the sludge can be a key indicator that something is wrong within the system, and operators should use the following table (Table 7.5) to recognize symptoms that the digester sludge might exhibit, which signal suboptimal operating conditions.

7.3 Volatile Sludge Troubleshooting

For any digestion system, careful control over the organic loading rate and environment inside the digester is critical for successful operation. This is especially true for co-digestion systems in which the organic content of the feedstock may be more variable. Operators may reference the following table (Table 7.6) to review how VS data can be used to diagnose operations issues. Table 7.6 also provides guidance for which corrective actions should be considered to recover from upset conditions.

7.4 Digester Gas Troubleshooting

Operating data from the digester gas system can provide operators with useful information not only about the gas system, but also about the health of the digester system. The following table (Table 7.7) can be used to help

TABLE 7.4 Thermal Hydrolysis Process System Process Control Troubleshooting

Condition	Possible Causes	Remedial Measures
THP Influent Total Solids Below 15%	• Poor pre-thickening equipment performance	• Improve pre-thickening equipment performance
THP Influent Total Solids Greater Than 17%	• Excessive solids detention time in the cake bin	• Lower and maintain the lower cake bin level
Insufficient pulper tank steam input (flow/temperature)	• Poor combined heat and power (CHP) system performance	• Correct the CHP system fuel source necessary to provide the steam

TABLE 7.5 Digester Sludge Troubleshooting

Indicator	Probable Cause	Check or Monitor	Potential Solution
Gray or brown color sludge	• Improper digestion • Insufficient mixing	• Check for layers of unmixed sludge in the bottom of the tank.	• Refer to mixing pump operating manual for troubleshooting instructions.
Sour odor	• Low pH • Second stage of digestion may be retarded • Overloaded digester	• Check pH at different levels. • Check CO_2 content of gas. • Check ratio of VS added to digester	• Add lime. • Pause digester loading and allow digester to rest.
High pH	• High alkalinity • Ammonia formation • Excessive foaming	• Check pH at different levels. • Check mixers.	• Increase mixing. • Add fresh water to reduce alkalinity. • Add acid. • Use antifoaming agents.
Digester full of well-digested sludge and supernatant suspended solids are high	• Not withdrawing enough digested sludge	• Check process indicators compared to normal values. • Check withdrawal records and compare to feed records.	• Increase digested sludge withdrawal to dewatering or disposal. Withdrawal rate should not exceed 5% of digester volume per day.
Complete lack of biological activity	• Highly toxic waste such as metals or bacteriocide	• Lack of gas production • Analyze sludge sample for presence of toxins.	• Empty contents or flush with thickened sludge feed for one complete tank volume and restart the biological feeding process.

TABLE 7.6 Volatile Solids Troubleshooting

Indicator	Probable Cause	Check or Monitor	Potential Solution
Rise in VA/ALK ratio due to hydraulic overload	• Overfeeding caused by accidental over pumping.	• Monitor volatile acids, alkalinity, and temperature twice daily until the problem is corrected. • If VA/ALK ratio rises to 0.5, the concentration of CO_2 in the gas starts to increase—monitor this using the waste gas burner and/or a gas analyzer. • If VA/ALK ratio rises to 0.8, the pH starts to drop, and CO_2 increases to the point that no burnable gas is obtained (42%–45%)—monitor gas conditions. Hydrogen sulfide or rancid butter odors may occur.	• If VA/ALK ratio increases to 0.3, decrease waste rate to keep feed sludge in digester and check sludge temperatures closely/control heating if necessary. • If VA/ALK ratio increases to 0.5, start adding alkali using the formulas as outlined in this manual. • If VA/ALK ratio increases to 0.8, add alkali using the formulas as outlined in this manual and decrease the loading rate to less than 0.01 lb vol. solids/cu ft/day until the VA/ALK ratio drops to 0.5 or less.
Rise in VA/ALK ratio due to organic overload	• Overfeeding caused by accidental overthickening of sludge feed.	• Monitor sludge pumping volume and amount of volatile solids in feed sludge. • If VA/ALK ratio rises to 0.5, the concentration of CO_2 in the gas starts to increase—monitor this using the waste gas burner and/or a gas analyzer. • If VA/ALK ratio rises to 0.8, the pH starts to drop, and CO_2 increases to the point that no burnable gas is obtained (42%–45%)—monitor gas conditions. Hydrogen sulfide or rancid butter odors may occur.	• If VA/ALK ratio increases to 0.3, decrease feed rate and check sludge temperatures closely/control heating if necessary. • If VA/ALK ratio increases to 0.5, start adding alkali using the formulas as outlined in this manual. • If VA/ALK ratio increases to 0.8, add alkali using the formulas as outlined in this manual and decrease the loading rate to less than 0.01 lb vol. solids/cu ft/day until the VA/ALK ratio drops to 0.5 or less.

TABLE 7.7 Digester Gas Troubleshooting

Indicator	Probable Cause	Check or Monitor	Potential Solution
Gas leaking through digester roof pressure relief valve	• Valve not seating properly or stuck open	• Check manometer for normal digester gas pressure	• Remove PRV cover and move weight holder until it seats properly. Install new ring if needed.
Digester gas operating pressure is higher than normal	• Obstruction or water in burner gas line. • Digester PRV is stuck shut. • Waste gas burner line pressure control valve is closed.	• If all other operating points are normal, check for a waste gas line restriction or a plugged or stuck safety device. • Gas is not escaping properly. • Gas meters indicate excess gas is being produced, but not going to waste gas burner.	• Purge with air; drain condensate traps and check for low spots. Do not force air into the digester. • Remove PRV cover and manually open valve; clean valve seat.
Digester gas operating pressure is lower than normal	• Withdrawal rate may be too fast, causing a reduction of pressure inside the digester • Excess alkali addition.	• Check vacuum breaker to verify it is operating properly. • Check for sudden increase in CO_2 in digester gas.	• Stop digester withdrawal and close off all gas outlets until pressure returns to normal. • Stop alkali addition.
Digester PRV is frozen	• Winter conditions	• Remove valve cover and inspect PRV.	• Apply mixture of salt and grease to seat ring. • Place vented barrel over valve with insulation and heat tape.

Pressure regulating valve not opening as pressure increases	• Inflexible diaphragm • Ruptured diaphragm	• Isolate valve and open cover.	• Use soap to identify leaks. • Ruptured diaphragm requires replacement.
Yellow gas flame from waste gas burner	• Poor quality gas with high CO_2 content.	• Check CO_2; content will be higher than normal.	• Check concentration of sludge feed; it may be too dilute. If so, increase volatile solids sludge concentration.
Gas flame lower than normal	• High gas usage • Gas leak from digester piping or safety devices • Low gas production due to process problems	• Check gas production rate against gas usage. • Check gas collection and distribution system, starting at digester main point.	• Check for escaping gas; if found, isolate and repair. • This may be a normal condition arising from supply/demand differences.
Waste gas burner not lit	• Pilot flame not burning. • Obstruction or water in main waste gas line or pilot natural gas line. • Pressure control valve closed.	• Check pilot line pressure at waste gas burner. • Check waste gas pressure control valve.	• Re-light natural gas pilot line. • Drain condensate traps. • Open valve and verify that setting will allow valve to open when pressure is about 1/4" water column above all other use point pressures.
Gas meter failure	• Debris in line • Electrical failure	• Check condition of gas line. • Check for fouled or worn parts.	• Flush with water, isolating digester and working from digester towards point of usage. • Replace worn parts.

operators to diagnose both mechanical and biological issues that can show up in the digester gas system and respond to them appropriately if/when they are identified.

7.5 Toxicity Troubleshooting

As a sensitive biological operating system, digesters are susceptible to toxic compounds that may inhibit biological activity. It is important to monitor feeds to digesters to avoid contamination with toxins, but the following table (Table 7.8) can be used to identify potential toxicity issues and respond depending on the probable type of toxin.

8.0 CRITICALITY REVIEW OF THE TREATMENT EQUIPMENT

No matter how effectively the operation and maintenance staff are monitoring the biomass and the daily organic loading rate, the process will degrade due to:

- Ineffective equipment
- Erratic equipment performance
- Equipment failure

This degradation will create a disturbance with the biomass efficiency defined as maximum biogas quality and volume.

8.1 Updating the Criticality Review

Criticality review is an activity that has probably a half-dozen formal and informal definitions. As a rule, most utilities fall into two major O&M performance categories:

- Run-to-failure
- Preventive maintenance

These two maintenance concepts address the facility, which may entail several thousand individual pieces of equipment, which may or may not be populated in a computerized maintenance management program.

While on the surface these methods appears to be an effective means to maintain the "Equipment Up-Time" requirements—defined as meeting

TABLE 7.8 Toxicity Troubleshooting

Indicator	Probable Cause	Check or Monitor	Potential Solution
• Heavy metals • VA/ALK increases • pH drops • Gas production decreases • Rancid butter odor (butyric acid)	• Industrial discharges	• Use atomic absorption to identify specific metal contaminant.	• Use any combination of the following: solids recycle, liquid dilution, decreased feed concentration, or heavy metal precipitation with a sulfur compound. Verify that pH in digester is greater than 7.
Sulfides	• Industrial discharges • Anaerobic activity in upstream processes	• Industrial wastes in raw influent • Inadequate sludge pumping	• Dilution • Precipitation using ferric chloride • Institute source control program for industrial wastes.
Ammonia	• Industrial discharges • Organic overload • Overcorrecting a pH problem	• Wastes from industrial users handling nitrogenous products • Check pounds volatile solids/day in feed.	• Dilution and solids recycle • Decrease feed concentration.
Excessive Alkali (sodium, potassium, calcium, magnesium)	• Industrial discharges • Overcorrecting a pH problem	• Wastes from industrial users handling industrial chemicals, fertilizers, etc. • Check alkalinity and specific constituent amounts.	• Use an antagonistic compound. • Alter digester volatile solids loading rate to 50% of "normal" feed rate.

a facility operating permit limitation, the efficiency of these methods are under a variety of constraints, including but not limited to:

- Budget
- Knowledgeable staff
- Quantity of staff
- In-house knowledge
- In-house knowledge transfer process
- Contractor availability
- Contractor capabilities

Preventive maintenance looks great on paper (or the computer screen) but attaining those criteria is rapidly becoming an impractical task. All equipment requires maintenance, but which equipment is critical for sustaining process up-time?

Running-to-failure (RtF) presents difficulties in these days of constrained staffing and financial resources, replacing equipment plus contending with continual "firefighting" and/or emergency work adds significant and unnecessary workplace stress and strain.

While on the surface both methods appear to be an effective means to maintain the "Equipment Up-Time" requirements, "Equipment Up-Time" effectiveness is instead defined for this section as the continual ability to maintain a facility operating permit limitation for both final effluent and biosolids quality. The authors' preferred maintenance philosophy is known as reliability-centered-maintenance (RCM)—a method that is focused on critical equipment that are required to attain/maintain the facility efficiency and effectiveness to meet or exceed the utility goals. This method aims to derive the benefits from both RtF and preventative maintenance methodologies. As part of RCM, criticality reviews are needed to assess process critical equipment and find potential points of failure.

The best period for the utility to perform a criticality review begins at the complete set of 30% P&ID (piping and instrumentation diagram). This discussion will assume the facility has been on-line for years and is in a reaction maintenance mode based on one or more of the issues listed above.

A criticality review will not resolve those issues, but a functional criticality review will allow the utility to make a coherent decision as to which specific pieces of process equipment (for example, heat exchanger water recirculation pump) and non-process equipment (for example, HVAC unit for a motor control center) to apply the limited resources available to maintain equipment up-time to ensure process biological stability and/or facility operating permit criteria.

8.2 How to Perform a Criticality Review

The facility critical review team should include process, mechanical, electrical, and instrumentation and control staff, working with a knowledgeable facilitator to trace through the "final" facility P&IDs to clarify the following:

- Single points of failure that impact immediate staff health and safety.
- Single points of failure that directly impact the ability to maintain process biological stability and/or operating permit compliance.
- Points of failure that have minimal facility impacts.

The criticality review should be performed, P&ID sheet by P&ID sheet, with notes that clarify the group determination of the criticality review statuses for each individual equipment, sub-system, and/or system. A component or system should be assigned a number status per the following:

- Number 1 status (immediate staff health and safety) that allows no immediate component replacement, and which will require both active maintenances to ensure continual operability and/or a reasonable SOP or workaround for the shift staff to execute.
- Number 2 status that allows at least one or no immediate component replacement, and which will require both active maintenance to ensure operability and/or a reasonable SOP or workaround, for the shift staff to execute.
- Number 3 status may require active maintenance to ensure operability plus a SOP or workaround for the shift staff to execute startup of the standby system.

Components listed with a status of 3 are now logically lumped into a basket of maintenance options:

- "Run-to-Failure" and replace as needed
- Perform equipment supplier maintenance "as-time/resources permits"
- Defer to contractor oversight
- Maintain as determined by predictive maintenance output data (oil analysis, vibration, thermal, or similar technologies)

There are numerous well-written and documented articles and books about criticality reviews. This work serves as a starting point for the reader to perform their own deep dive on the subject.

9.0 REFERENCES

Gerardi, M. (2019). *Wastewater odor production and control.* DEStech Publications, Inc.

Gerardi, M. H. (2003). *The microbiology of anerobic digesters.* John Wiley & Sons, Inc.

McKinney, R. (1962). *Microbiology for sanitary engineers.* McGraw-Hill.

Metcalf & Eddy, Inc./AECOM. (2014). *Wastewater engineering treatment and resource recovery* (5th ed.). McGraw-Hill.

Speece, R. E. (1996). *Anaerobic biotechnology for industiral wastewaters.* Archea Press.

Speece, R. E. (2008). *Anaerobic biotechnology and odor control: Corrosion control for municpalites and industries.* Archea Press.

Water Pollution Control Federation. (1987). *Anaerobic sludge digestion.*

10.0 SUGGESTED READINGS

Carlson, C. S. (2012). *Effective FMEA's achieving safe, reliable, and economical products and process using failure mode and effects analysis.* John Wiley & Sons, Inc.

Gerardi, M. H. (2003). *The microbiology of anaerobic digesters.* Wiley-Interscience.

Gulati, R. (2013). *Maintenance and reliability best practices* (2nd ed.). Industrial Press, Inc.

Levy, P. F. (2001). *The Nut Island effect: When good teams go wrong.* Harvard Business School Publishing.

McCarty, B. R. (2001). *Environmental biotechnology principles and applications.* McGraw-Hill.

Moubray, J. (1997). *Reliability-centered maintenance* (2nd ed.). Industrial Press, Inc.

Palmer, D. (2013). *Maintenance planning and scheduling handbook* (3rd ed.). McGraw-Hill.

Speece, R. E. (2008). *Anaerobic biotechnology and odor/corrosion control for municipalities and industries.* Archaea Press.

Water Environment Federation. (2023). *Solids process design and management* (2nd ed.).

Chapter 8

Biogas and Bioproducts Production and Utilization From Co-digestion

Juliet Ohemeng-Ntiamoah, PhD, PE; Temesgen Fitamo, PhD, PE; Pedro Puente; & Arifur Rahman, PhD, PE

1.0 FACTORS AFFECTING BIOGAS PRODUCTION

Anaerobic co-digestion is implemented to mainly augment biogas (and other bioproducts) production. However, as discussed in Chapters 3, 4, 6, and 7, there are factors that can inhibit or enhance the co-digestion process and biogas production. Some of the factors that affect biogas production during co-digestion are briefly discussed in the subsections below.

1.1 Co-substrate Thermal and Microbial-Enabled Hydrolysis

Most co-substrates are typically easily biodegradable (e.g., fats, oil, and grease [FOG] or food waste) and substantially increase biogas production without pretreatment. Research has shown that thermal hydrolysis pretreatment of food waste provides marginal additional increase in biogas production (Ismail et al., 2022). However, thermal and microbial hydrolysis pretreatment of sludge that is co-digested can further improve biogas production in the co-digestion process. Microbial hydrolysis pretreatment of cellulosic waste and lignocellulosic wastes such as yard trimmings can improve biogas production in anaerobic digestion and co-digestion processes (Parry et al., 2022; Pittman et al., 2022).

1.2 Co-substrate Fractions

The optimal co-substrate fraction for the highest biogas production can vary between water resource recovery facilities (WRRFs) and is dependent on organic loading rate. However, co-substrates fractions below 40% of the volatile solids loading can significantly increase biogas production without causing digester upsets. Most studies have found gradual increase of co-substrate fraction can be used to induce microbial acclimation for high biogas production at high co-substrate fractions without causing digester upsets (Ohemeng-Ntiamoah, 2020).

1.3 Alkalinity/Volatile Fatty Acids/pH

The impact of alkalinity, volatile fatty acids (VFAs), and pH are discussed in Chapters 3, 6, and 7. The necessary pH for stable anaerobic co-digestion ranges between 6.8 to 7.2. Most co-substrates such as FOG and other food waste have lower pH. Digesters with enough alkalinity will be able to function well and produce stable biogas production even with lower pH co-substrates if loaded at a reasonable rate. Accumulation of VFAs may also occur during co-digestion if the rate of VFA production by acidogens exceeds the rate of its consumption by methanogens due to the high degradability of co-substrates. In such instances, sufficient alkalinity is needed to neutralize the VFAs produced to maintain stable biogas production.

1.4 C:N ratio

As discussed in Chapter 3, municipal sludge has lower C:N compared to typical co-substrates. The lower C:N has potential to increase ammonia toxicity while high C:N increases the likelihood of volatile fatty acid accumulation. Combination of municipal sludge and carbohydrate-rich co-substrates such as organic fraction of municipal solid wastes creates more balanced C:N for enhanced biogas production in anaerobic co-digestion systems. A C:N of 20–30:1 should be targeted for optimal biogas production.

1.5 Hydraulic Retention Time and Organic Loading Rate

Similar to conventional anaerobic digestion systems, the methanogens responsible for biogas production require longer hydraulic retention time (HRT) ($\gtrsim$15 days) to grow and function efficiently to produce biogas. Shorter HRT can result in the washout of the methanogens and reduction in biogas production in co-digestion systems. Organic loading rate (OLR) has a strong relationship with biogas production in that biogas production increases with organic loading rate until the digester's capacity threshold is exceeded. The impact of HRT and OLR on co-digestion are discussed in more detail in Chapters 6, 7, and Section 2.0 below.

2.0 BIOGAS AND BIOPRODUCTS PRODUCTION THROUGH CO-DIGESTION

This section discusses the potential of co-digestion in WRRFs, optimization of the anaerobic co-digestion process, and high-value-added bioproduct production through anaerobic co-digestion.

2.1 Co-digestion of Sludge With High-Strength Waste

Water resource recovery facilities may have excess digester capacity that can be used with minor modifications to process high-strength waste (HSW) like food waste and FOG to produce extra biogas, reduce sludge volume, and stabilize the waste. Large quantities of HSW are generated in populous metropolitan areas, making co-digestion with WRRF sludge a sustainable alternative to landfilling (U.S. EPA, 2023). Examples of co-digestion implementation at WRRFs are provided below and in Chapter 9.

2.2 Optimization of Anaerobic Co-digestion of Co-Substrates

The anaerobic co-digestion process must be optimized to achieve maximum gas production and enhance solids destruction efficiency, while minimizing VFAs and ammonia accumulation to avoid digester upsets (Fitamo et al., 2016). Maintaining nutrient balance and microbial communities is critical for achieving these objectives, with OLR and HRT as key operational parameters.

Organic loading rate can be increased to boost gas production and methane yield, but excessive loadings can cause digester upsets. The recommended maximum volatile solids (VS) and chemical oxygen demand (COD) loadings are 0.2 lb/day/cu ft (3.2 kg/day m^3) (Water Environment Federation, 2021) and 0.249 lb/day/cu ft (4 kg/day/m^3), respectively (Metcalf & Eddy, Inc./AECOM, 2014). In addition, Angelidaki et al. (2009) recommends a co-substrate ratio of approximately 0.5 lb COD or VS feed per lb VS in digester, to maintain a healthy, balanced microbial community and maximize biogas production.

Hydraulic retention times must be balanced for methane yield, microbial washout, regulatory requirements, and digester capacity. Shorter HRTs allow higher throughput and capacity, but at values less than 10 days (Fitamo et al., 2017; Zhang et al., 2023) critical microbes can wash out, resulting in digester upsets. In addition, some WRRFs operate with a minimum residence time of 15 days to satisfy the U.S. EPA Part 503 rule for anaerobic digestion of Class B biosolids. Higher HRTs result in higher methane yields but necessitate larger digester capacity or reduce the allowable throughput in existing facilities.

2.3 Production of Bioproducts and Nutrient Recovery Through Co-digestion of Organic Wastes

Anaerobic co-digestion produces a nutrient-rich byproduct called biosolids that can be used as a biofertilizer for agricultural soils. The biosolids from the co-digestion of sewage sludge with food waste have nitrogen, phosphorus, and potassium, making it suitable for soil amendment (Akbar et al., 2021). The biosolids can be applied on land as a biofertilizer to improve soil fertility and crop productivity; however, new upcoming per- and poly-fluoroalkl substances (PFAS) regulations must be considered.

Volatile fatty acids, such as acetic acid, propionic acid, and butyric acid, produced through co-digestion, can be used to create a variety of chemicals and bio-based products (Wang et al., 2023). These biochemicals have potential applications in bioplastics, biofuels, and sustainable chemical synthesis. The bio-chemical production from co-digestion is still in the research and development stage. Start-ups like Bioveritas, Chaincraft, and Capro-X are converting waste into valuable products. Chaincraft, based in the Netherlands, operates a commercial-scale facility that converts food waste into medium chain carboxylic acid (MCCAs) and alcohol. Bioveritas, located in Texas, is scaling up from pilot-scale systems to commercial facilities. Capro-X, a newer company, focuses on MCCAs production from acid whey at the pilot scale. The process of anaerobic co-digestion shows a great potential for converting organic waste materials into valuable bioproducts and replacing petrochemicals.

3.0 IMPACT OF CO-DIGESTION ON BIOGAS PRODUCTION AND QUALITY

The implementation of co-digestion has been studied at bench and pilot scales with the ultimate goal of applying it at full-scale facilities. At bench scale, the study of co-digestion research focuses on identifying the maximum gas production from substrates using biomethane potential tests. Pilot-scale tests are conducted to evaluate specific substrates and operational conditions that WRRFs may use. Pilot studies also help anticipate potential challenges WRRF operators may encounter during broader co-digestion application such as pumping material, dosing of alkalinity, handling of increased gas production, and digested sludge management.

The cases presented below in Table 8.1 summarize the experience of U.S. WRRFs with co-digestion of sewage sludge and other organic streams (Dalke et al., 2021; U.S. EPA, 2023). Of the 16,000 WRRFs in the U.S., between 80 and 200 have implemented co-digestion to enhance energy recovery (Rauch-Williams et al., 2022). Commonly used co-substrates include FOG, HSW which consists of local food processing waste, and food waste (FW) from restaurants, hospitals, and so on. The literature review showed that co-digestion can boost biogas production between 20% and 100%.

Co-digestion results in an important increase in biogas production at WRRFs compared to sludge mono-digestion. However, studies indicate that methane content of biogas sometimes remain relatively unaffected in sludge

TABLE 8.1 Water Resource Recovery Facilities Co-digesting Sewage Sludge and Organic Waste Streams in the United States

Project name	Location	Co-substrate flow	Biogas production	Biogas use
City of Flagstaff	Flagstaff, AZ	15.12-m^3 FOG/d (4000-gal/d)	1456 m^3/d (51 420 cu ft/d)	CHP (when operational) or flared
Carmel Area Wastewater District WWTP	Carmel, CA	15.12-m^3 FW/week (4000 gal/week)	991 m^3/d (35 000 cu ft/d)	CHP: microturbines produce 65–80 kW electricity; waste heat used for digesters
Central Marin Sanitation Agency (CMSA)	San Rafael, CA	5.67 tonnes FW/d (6.25 tons/d) 51.03 m^3 FOG/d (13 500 gal/d)	7900 m^3/d (279 000 cu ft/d)	CHP: 750-kW engine generates heat and power for internal use and CMSA sells excess electricity
Encina Wastewater Authority	Carlsbad, CA	32.87 m^3 FOG/d (8700 gal/d)	20 470 Sm^3/d (722 900 scf/d)	Four Caterpillar turbo charged engines can generate 750 kW. Heat is used in digesters. 80% of facility electricity demand is covered by biogas.
Fresno Clovis RWRF	Fresno, CA	29 106 m^3 FOG and HSW/yr (7.7 MG/yr)	39 644 m^3/d (1 400 000 cu ft/d)	Biogas can be used in two 3.3 MW gas turbines and a 2.2 MW steam turbine, a 400 HP boiler or be flared. In 2015, they produced 21 261 MWh of power between biogas and natural gas.
Hill Canyon Treatment Plant	Camarillo, CA	94.50 m^3 FOG/d (25 000 gal/d) 94.50 gal biodiesel waste/d (25 000 gal/d)	12 743 m^3/d (450 000 cu ft/d)	CHP: 1120 kW capacity

A.K. Warren Water Resource Facility	Carson, CA	59–77 tonnes FW/d (65–85 tons/d)	206 713 sm^3/d (7 300 000 scf/d/d)	CHP via three 9.9 MW turbines for internal usage and exporting
Urbana and Champaign Sanitary District	Urbana, IL	19 354 m^3 FOG and FW/yr (5.12 MG/yr)	2 208 710 m^3/yr (78 000 000 cu ft/d/yr)	CHP: 2.644 MWh/yr covering 30% of facility's power demand
West Lafayette WRRF	West Lafayette, IN	0.54 m^3 FOG and FW/d (143 gal /d)	2610 m^3/d (92 160 cu ft/d/d)	CHP: two 65 kW microturbines generate electricity and heat for digesters; 679 MWh/yr electricity
City of St. Cloud NEW Recovery Facility	St. Cloud, MN	98.28 m^3 HSW/d (26 000 gal/d)	9911–11 327 m^3/d (350 000–400 000 cu ft/d)	Two 633 kW Jenbacher Internal Combustion Engine/Generator. With appropriate conditions, the facility can operate energy neutrally. HSW addition doubled their biogas production.
Rahway Valley Sewerage Authority	Rahway, NJ	Capacity for 756 m3 FW/week (200 000 gal/week)	7079 m^3/d before co-digestion (250 000 cu ft/d/d)	Four 1.55 MW CHP engines. Biogas production expected to double with co-digestion
NYCDEP Newtown Creek Wastewater Resource Recovery Facility Codigestion Program	Brooklyn, NY	118 tonnes FW/d (130 tons/d)	38 794 m^3/d (1 370 000 cu ft/d)	Biogas heats boilers; excess supplied to the grid; has partnered with National Grid to develop RNG
Wooster WWTP	Wooster, OH	226.8 m3 FOG/d (60 000 gal/d)	9895 m^3/d (349 440 cu ft/d/d)	CHP: 1100 kW generator
Fond du Lac WTRRF	Fond du Lac, WI	75.60–94.50 m^3 HSW/d (20 000–25 000 gal/d)	5663 m^3/d (200 000 cu ft/d/d)	CHP: 450 kW engine

co-digestion as seen in Figure 8.1 (Subramanian et al., 2022). Some co-substrates increase methane content of biogas, while others decrease it, and currently there is no general rule to predict these outcomes.

4.0 IMPACTS OF CO-DIGESTION ON BIOGAS HANDLING AND UTILIZATION

4.1 Impact of Co-digestion on Biogas Handling Equipment and Treatment

The biogas handling and treatment for co-digestion systems is similar to conventional anaerobic digestion systems except co-digestion systems will require appropriately sized equipment to accommodate the increase in biogas production as discussed in Chapter 6. In addition, some co-substrates can impact biogas quality by increasing the concentration of hydrogen sulfide (H_2S), volatile organic compounds and ammonia. For example, co-digestion

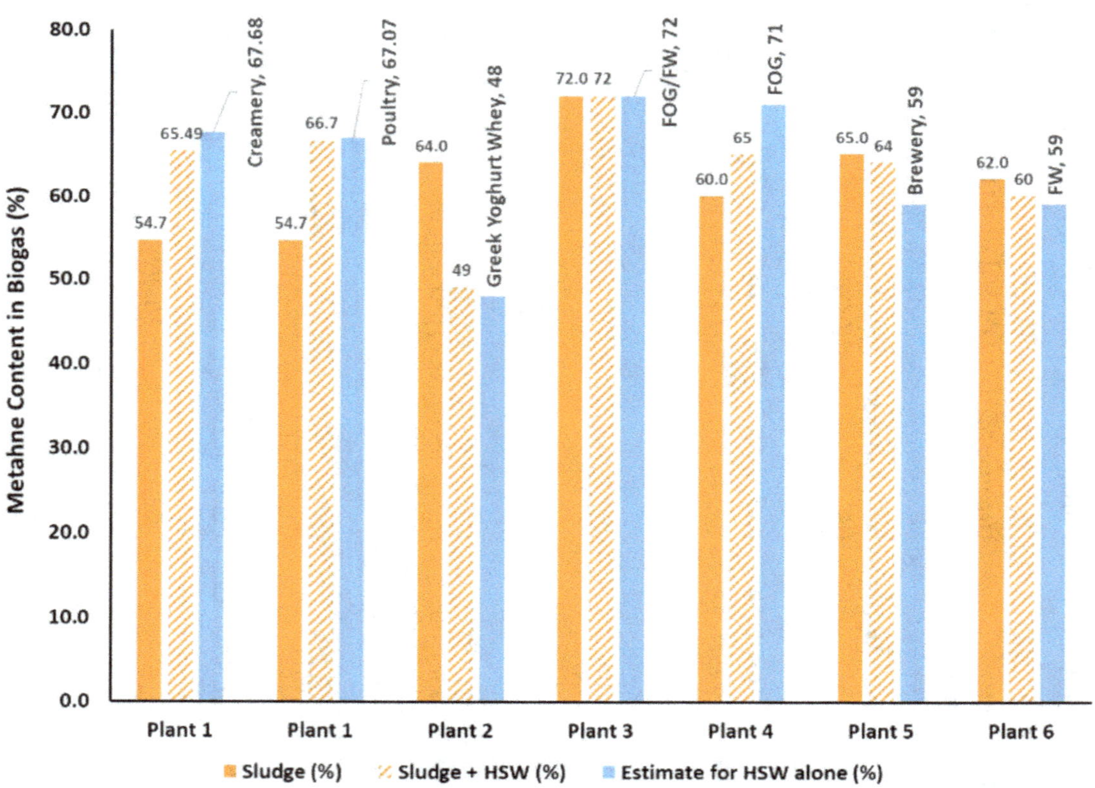

FIGURE 8.1 Impact of Co-digestion With High-Strength Waste on Methane Content (Reprinted from Subramanian et al., 2022)

of poultry waste with sludge has been shown to increase H_2S concentration from <100 ppmv to ~600 ppmv (Subramanian et al., 2022). The application of the biogas treatment process depends on raw biogas composition and biogas utilization. The biogas treatment process can involve removing H_2S, moisture, and siloxanes with iron sponges, chillers and granular activated carbon filters for utilization in combined heat and power (CHP) systems or treatment to pipeline gas quality.

4.2 General Biogas Utilization

This section explains the biogas usage in difference pathways including CHP system, renewable natural gas (RNG) production and its injection to gas utility grid system, and biogas onsite usage. A more comprehensive discussion of potential biogas usage pathways can be found in Chapter 10 "Digester Gas to Energy Systems" in WEF Manual of Practice No. 32, *Energy in Water Resource Recovery Facilities* (2021).

4.2.1 Biogas to On-site Power Generation and Heat via Combined Heat and Power

Because RNG pipeline injection and the market demand has uncertainties, utilities are looking into other options for renewable energy production. Converting biogas to electricity and heat via CHP systems is another method of renewable energy production and many utilities are practicing this technology in the United States. The process involves biogas cleaning to remove moisture, siloxane, and H_2S, followed by compression and feed to a CHP engine to generate electricity and heat simultaneously. The high-grade heat recovered from the CHP can be used for steam boiler to run thermal hydrolysis process (THP) process. The low-grade heat from the CHP can provide heating requirements for building heating (non-process heating). Without THP, heat recovered from CHP can be used for hot water boiler for anaerobic digester and non-process heating (building). Several types of CHP prime movers are available, but spark ignition reciprocating internal combustion engine is most used for biogas. The renewable electricity produced from CHP can be used for electric vehicle charging station or electrical grid injection. The CHP process can provide resilient power 24 hours a day, 7 days a week in the event of grid outages, and it can be paired with distributed energy technologies such as solar photovoltaics, and energy storage.

The future economics of CHP systems can become better if electric Renewable Identification Number (eRINs) are approved. The eRINs program will be similar to the RINs program described in Section 4.2.5 except that eRINs will be applicable to the electric vehicle market. Renewable identification numbers are unique identification numbers that are assigned to batches of renewable energy produced and released into the grid. These RINs can be traded to generate revenue for the producer. Producers of electricity generated from renewable fuel sources (e.g., biogas) will be able to generate revenue from the electricity when it is supplied to the grid for use in electric vehicles. The eRINs program by the U.S. Environmental Protection Agency (U.S. EPA) remains uncertain and is not yet finalized as of this time of this publication.

The Partnership for Electric Pathways was formed in November of 2023 to ensure the U.S. EPA finalizes its Renewable Fuel Standard (RFS) "Electric Pathways" framework proposal as soon as possible. The RFS Electric Pathways are fundamentally important to enabling our vision of a future in which electric vehicles can be powered by organic waste. Planners and designers should check for updates on eRIN standards because they will increase the economic viability of energy recovery from biogas.

4.2.2 Biogas On-site Utilization

4.2.2.1 Conditioned Biogas for Boiler and Thermal Dryer Use

Biogas typically serves as a common primary fuel source for hot water boilers, THP steam boilers, and thermal dryers. These boilers are typically designed to operate using dual fuel sources: conditioned biogas from anaerobic digesters (after H_2S removal) as the primary fuel and other sources such as natural gas (NG) or liquefied natural gas (LNG). When continuous biogas supply from the digester is unavailable, temporary biogas storage becomes necessary (Figure 8.2).

4.2.2.2 Conditioned Biogas Storage at Low and High Pressure

Biogas storage is essential to maintain a consistent feed rate to the CHP engine and enhance operational reliability. During the design phase, the size of the biogas storage is carefully adjusted, considering the daily fluctuations in biogas production and the capacity of the CHP system. Biogas storage can also provide fuel supply to boilers when production is not sufficient.

Conditioned biogas can be stored at low pressure (<2 psi) when production is sufficient for short-term use. Low-pressure storage systems maintain constant pressure with variable volume, allowing for expansion during storage needs and contraction during gas depletion. Conditioned biogas storage at low pressure requires larger volume. Alternatively, conditioned biogas can be compressed at high pressure (3000–5000 psi) and stored in steel or alloy steel tanks (spherical or cylindrical). Compressed biogas has a useful life of 15 to 20 years, depending on tank construction. Microturbines, for example use biogas stored at medium pressure range (5–100 psi). When more than 10 000 pounds of methane is stored onsite, the Occupational Safety and

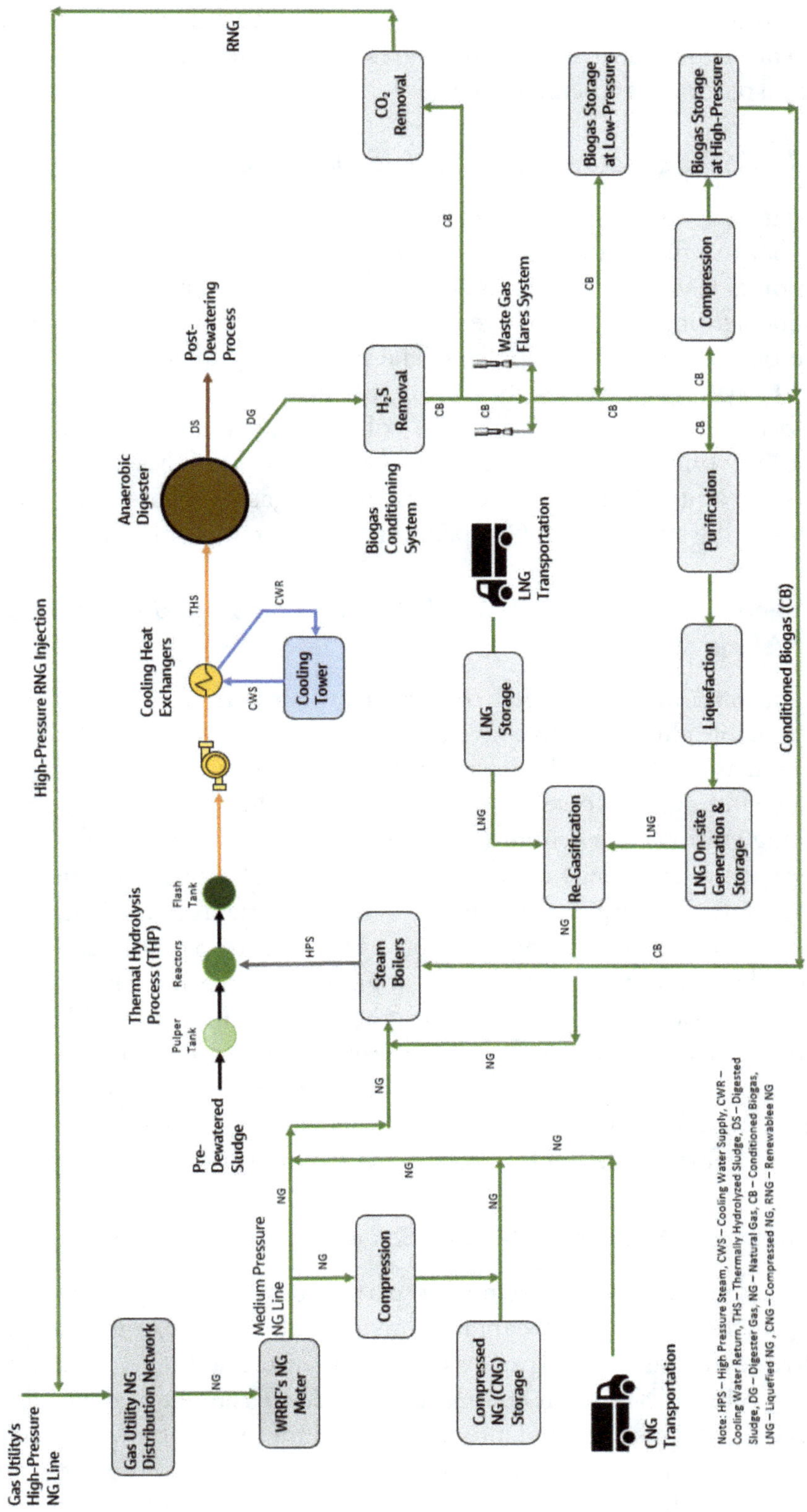

FIGURE 8.2 Process Flow Diagram of Biogas Utilization Via Storage, Liquified Natural Gas, and Renewable Natural Gas Injection

Health Administration's (OSHA, 2023) regulations require that facilities implement comprehensive safety measures, such as risk management plans, regular inspections, and employee training.

4.2.2.3 Conditioned Biogas to Liquified Natural Gas

Biogas can be used to produce liquified natural gas (LNG), which serves as a reliable backup fuel option for THP steam boilers (Figure 8.2). Onsite LNG generation is feasible, involving several complex stages, including purification, liquefaction, storage, and re-gasification. The approach helps reduce reliance on external LNG suppliers. The process of liquefaction cools the cleaned biogas to an extremely low temperature of −260 °F (−162 °C), resulting in a significant reduction in volume (approximately 1/600th of the original volume). A regasification system, also known as a vaporizer, converts the liquid biogas (or NG) back into its gaseous state. The storage volume and duration for LNG depend on specific requirements.

4.2.3 Biogas to Renewable Natural Gas and Compressed Natural Gas Vehicle

Renewable natural gas, derived from organic waste materials, holds promise as a sustainable alternative to conventional natural gas. When purified, RNG can be used as a vehicle fuel, offering reduced emissions and contributing to environmental goals. However, establishing dedicated compressed natural gas fueling stations is essential for its successful adoption. These stations vary in size and cost, depending on factors such as fleet size, location, and operational needs. Despite the challenges, RNG's benefits make it an attractive choice for a greener transportation future. Fleet operators can power buses, trucks, and other fleet vehicles using RNG. Partnering with transportation companies allows them to reduce their carbon footprint and comply with environmental regulations. Local governments can integrate RNG into their public transportation systems, garbage trucks, and other municipal vehicles. Companies with large vehicle fleets (such as delivery services or waste management) can benefit from RNG partnerships.

4.2.4 Biogas to Renewable Natural Gas and Pipeline Injection

Biogas purification involves removing carbon dioxide (CO_2) and various pollutants (such as H_2S, mercaptans, organic sulfides, siloxanes, and aromatic hydrocarbons), as well as moisture. The resulting fuel is comparable in heating value to NG and is commonly referred to as biomethane or RNG. This RNG can then be injected into the NG grid for distribution. Gas quality requirements for injecting into the gas utility pipeline are necessary (typically >95% methane content). The injection process entails constructing compressors and a new high-pressure line connecting the process facility to the gas utility's high-pressure distribution system. Water resource recovery facilities can leverage this new injection line as a two-way conduit, using it as an alternative source of fuel for their processes when needed. Renewable natural gas producers can collaborate with a wide range of end users including industrial facililties, fleet operators, municipalities, and private and individual consumers.

4.2.5 Renewable Identification Numbers and Low Carbon Fuel Standard

When the RNG injected into the gas pipeline is used as transportation fuel, it qualifies to generate "renewable identification number" (RINs). Renewable identification numbers are unique identification numbers that can be traded to generate revenue for the RNG producer. The value of RINs depends on the type of feedstock used to produce the RNG and the volatility in the RINs market. Renewable natural gas produced from digesting cellulosic material such as wastewater sludge is classified as D3 RINs, which has the highest market value with a recent historical 5-year average of $2/RIN (per U.S. EPA's website updated August 10, 2023). Previously, all RNG produced in co-digestion systems were categorized as D5 RINs, which has a lesser market value. In a recent EPA set rule finalized in June 2023, the U.S. EPA made changes to include RIN apportionment between D3 and D5, RINs whereby biogas produced from co-digestion systems can be apportioned as both D3 and D5, for the biogas derived from sludge and the co-substrate, respectively. Water resource recovery facilities may need historical biogas production data (before implementing co-digestion) for the sludge-only digestion to be able to calculate the D3 component of the biogas produced from the co-digestion system.

Aside from the U.S. EPA RINs program, local states such as California have the Low Carbon Fuel Standard (LCFS), which are essential tools in promoting renewable energy and reducing greenhouse gas emissions. The LCFS, implemented in California, aims to lower the carbon intensity of transportation fuels by setting annual targets. Fuel producers can generate LCFS credits by producing fuels with lower carbon intensities than the set targets, which can then be traded in the market (U.S. EPA, 2018). Both LCFS credits and RINs provide economic incentives for the production and use of cleaner fuels, thereby supporting the transition to a more sustainable energy landscape (U.S. EPA, 2018).

5.0 IMPACT OF CO-DIGESTION ON GREENHOUSE GAS EMISSIONS AND CARBON FOOTPRINT

There is an inverse relationship between renewable energy production/utilization and greenhouse gas emissions. Thus, when WRRFs incorporate co-digestion in their systems, there is an indirect benefit of reducing their carbon footprint through replacement of fossil fuels and or diversion of organic waste from landfills.

5.1 Carbon Footprint Reduction Through Biogas Production and Utilization

The prime driver of anaerobic co-digestion is to boost biogas production as a renewable energy source for beneficial use and/or revenue generation. As discussed in Section 4.0, the use of biogas produced at WRRFs typically reduces and sometimes replaces the dependence on fossil fuels for heating and electricity. The reduction in the use of fossil fuels also translates into a reduction in greenhouse gases. For instance, a large WRRF that uses natural gas to thermally dry biosolids to Class A pellets can reduce its green house gas (GHG) emissions by at least 50% if biogas is used for thermal drying instead of natural gas. Thus, co-digestion presents a pathway for WRRFs to meet their decarbonization goals.

In addition to emission reduction through utilization of internally generated renewable energy, a recent life cycle assessment study conducted by Karolinczak et al. (2024) has showed that less GHGs are emitted during the production of one cubic meter (1 m^3) of biogas in a co-digestion system than from mono-digestion of wastewater residuals as shown in Figure 8.3. This could be attributed to the energy used in upstream wastewater treatment that generates wastewater residuals that is higher than the energy needed in the pretreatment of co-substrates for co-digestion. Thus, 1 kwh of energy

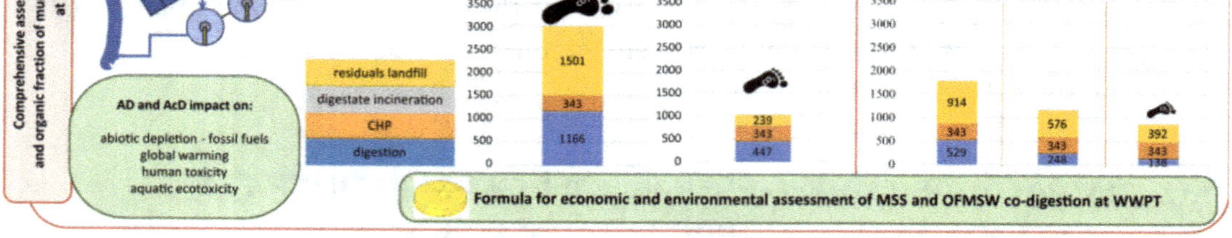

FIGURE 8.3 Carbon Footprint Comparison of Mono-digestion and Co-digestion Systems (Karolinczak et al., 2024). Licensed under CC BY 4.0.

produced from mono-digestion of sludge can have higher carbon footprint than 1 kwh of energy produced from a co-digestion system.

5.2 Carbon Footprint Reduction Through Diversion of Organic Waste From Landfills

There are many states such as California and New York that have enacted regulations regarding reduction of organic waste and diversion of organic wastes from landfills to more beneficial uses such as composting and anaerobic co-digestion systems. The diversion of organic waste from landfills prevents the direct emissions of GHGs into the atmosphere. The emissions avoided by uncontrolled emissions from organic waste in the landfills can result in an overall co-digestion emission reduction factor of 0.65–0.70 metric tons of carbon dioxide equivalents per wet short ton of diverted organic waste (MT CO2eq/wet short ton of diverted food waste) depending on the energy end use (Gupta, 2020).

6.0 REFERENCES

Akbar, S., Ahmed, S., Khan, S., & Badshah, M. (2021). Anaerobic digestate: A sustainable source of bio-fertilizer. In M. K. Jhariya, A. Banerjee, R. S. Meena, S. Kumar, A. Raj (Eds.), *Sustainable intensification for agroecosystem services and management*. Springer. https://doi.org/10.1007/978-981-16-3207-5_15

Angelidaki, I., Alves, M., Bolzonella, D., Borzacconi, L., Campos, J. L., Guwy, A. J., Kalyuzhnyi, S., Jenicek, P., & van Lier, J. B. (2009). Defining the biomethane potential (BMP) of solid organic wastes and energy crops: A proposed protocol for batch assays. *Water Science and Technology, 59*(5), 927–934. doi: 10.2166/wst.2009.040. PMID: 19273891.

Dalke, R., Demro, D., Khalid, Y., Wu, H., & Urgun-Demirtas, M. (2021). Current status of anaerobic digestion of food waste in the United States. *Renewable and Sustainable Energy Reviews, 151*(July), 111554. https://doi.org/10.1016/j.rser.2021.111554

Fitamo, T. M., Boldrin, A., Dorini, G., Boe, K., Angelidaki, I., & Scheutz, C. (2016). Optimising the anaerobic co-digestion of urban organic waste using dynamic bioconversion mathematical modelling. *Water Research, 106*, 283–294.

Fitamo, T., Treu, L., Boldrin, A., Sartori, C., Angelidaki, I., & Scheutz, C. (2017). Microbial population dynamics in urban organic waste anaerobic co-digestion with mixed sludge during a change in feedstock composition and different hydraulic retention times. *Water Research, 118*, 261–271.

Gupta, R. (2020). Comprehensive assessment of co-digestion capacity, costs, impacts, and operational history in California. *Proceedings of the 93rd Annual Water Environment Federation Technical Exhibition and Conference*, October 5-9, Water Environment Federation.

Ismail, A., laqa Kakar, F., Elbeshbishy, E. & Nakhla, G. (2022). Combined thermal hydrolysis pretreatment and anaerobic co-digestion of waste activated sludge and food waste. *Renewable Energy*, *195*, 528–539.

Karolinczak, B., Walczak, J., Bogacka, M., & Zubrowska-Sudol, M. (2024). Life cycle assessment of sewage sludge mono-digestion and co-digestion with the organic fraction of municipal solid waste at a wastewater treatment plant. *Science of the Total Environment, 907*, 167801.

Metcalf & Eddy, Inc./AECOM. (2014). *Wastewater engineering: Treatment and resource recovery* (5th ed.). McGraw Hill.

Ohemeng-Ntiamoah, J. (2020). *Anaerobic co-digestion of waste activated sludge with food waste and fats, oils and grease: Effects on digester performance, microbial community structure and activity* [Doctoral dissertation, Tennessee Technological University].

Parry, D., Fairley-Wax, M., Klibert, C., McClelland, S., Johnston, A., & Daniels, M. (2022, October). Enhancing anaerobic digestion performance with the microbial hydrolysis process. *Proceedings of the 95th Annual Water Environment Federation Technical Exhibition and Conference*, New Orleans, LA, October 8-12, Water Environment Federation.

Pittman, S., Young, M., & Ritmmann, B. (2022). *More methane and less landfilling? The case for anaerobic co-digestion of yard trimmings and wastewater*. Residuals and Biosolids 2022. Water Environment Federation.

Rauch-Williams, T., Van Steendam, C., Puente, P., Raskin, L., Steiniger, B., Schwinghammer, L., & Schaum, C. (2022). *Characterization of source separated food waste for co-digestion in water resource recovery facilities* (Issue WRF4915).

Subramanian, B., Rajagopalan, G., Higgins, M., Beightol, S. (2022). Characterization of biogas derived from wastewater solids and co-digested organic wastes. *Proceedings of the 95th Annual Water Environment Federation Technical Exhibition and Conference*, New Orleans, LA, October 8-12, Water Environment Federation. https://www.accesswater.org/?id=-10083783

U.S. Environmental Protection Agency. (2018). *Legal pitfalls when generating RINs and LCFS credits on your biogas*. https://www.epa.gov/sites/default/files/2018-03/documents/4_dave_mccullough_lmop.pdf

U.S. Environmental Protection Agency. (2023). *Renewable fuel standard (RFS) program: Standards for 2023–2025 and other changes*. https://www.govinfo.gov/content/pkg/FR-2023-07-12/pdf/2023-13462.pdf

U.S. Environmental Protection Agency, & Schroeder, J. (2023). *Anaerobic digestion facilities processing food waste in the United States* (2019). https://www.epa.gov/anaerobic-digestion/anaerobic-digestion-data-collection-project#results

U.S. Occupational and Health Administration. (2023). *Applicability of PSM standard to on-site use of methane as fuel.* https://www.osha.gov/laws-regs/standardinterpretations/2000-09-21

Wang, Y., Zhang, Z., Wang, X., Guo, H., Zhu, T., Zhao, Y., Lu, X., Zhang, Y., Ni, B.-J., & Liu, Y. (2023, December 1). Medium-chain fatty acids production from sewage sludge through anaerobic fermentation: A critical review. *Chemical Engineering Journal*, *477*, 147138.

Water Environment Federation. (2021). Digester gas to energy systems. In *Energy in water resource recovery facilities* (Manual of Practice No. 32).

Zhang, X., Jiao, P., Zhang, M., Wu, P., Zhang, Y., Wang, Y., Xu, K., Yu, J., & Ma, L. (2023). Impacts of organic loading rate and hydraulic retention time on organics degradation, interspecies interactions and functional traits in thermophilic anaerobic co-digestion of food waste and sewage sludge. *Bioresource Technology*, *370*, 128578.

7.0 SUGGESTED READINGS

Boldrin, A., Baral, K. R., Fitamo, T. M., Vazifehkhoran, A. H., Jensen, I. G., Kjærgaard, I., Lyng, K-A., Van Nguyen, Q., Nielsen, L. S., & Triolo, J. M. (2016). Optimised biogas production from the co-digestion of sugar beet with pig slurry: Integrating energy, GHG and economic accounting. *Energy*, *112*, 606–617.

Fuentes, A. (2023). *Bioenergy gas upgrading system, biogas to RNG workshop*. Water Research Foundation.

Gómez-Quiroga, X., Aboudi, K., Fernández-Güelfo, L. A., Álvarez-Gallego, C. J., &Romero-García, L. I. (2021). *Thermophilic anaerobic co-digestion of exhausted sugar beet pulp with cow manure to boost the performance of the process: The effect of manure proportion, water.*

Ma, G., Ndegwa, P., Harrison, J. H., & Chen, Y. (2020). Methane yields during anaerobic co-digestion of animal manure with other feedstocks: A meta-analysis. *Science of the Total Environment*, *728*, 138224. https://doi.org/10.1016/j.scitotenv.2020.13822

U. S. Environmental Protection Agency. (2021). *Anaerobic digestion facilities processing food waste in the United States (2017 & 2018).*

Water Research Foundation. (2023). *Biogas to RNG workshop.*

Yan, Y., Du, Z., Zhang, L., Feng, L., Sun, D., Dang, Y., Holmes, D. E., & Smith, J. A. (2019). *Identification of parameters needed for optimal anaerobic co-digestion of chicken manure and corn stover, RSC Advances.*

Yu, X., Wang, Y., Yan, L., Wang, Y., Lu, J., Huang, Y., Bi, S., & Wang, W. (2023). *Synergistic effects of anaerobic co-digestion of chicken manure and corn stover in batch and continuous modes, fermentation.*

Zhang, W., Werner, J. J., Agler, M. T., Angenent, L. T., & Chen, S. (2018). Strain improvement of anaerobic fungus anaeromyces robustus by adaptation to high ammonia concentration. *Biotechnology and Bioengineering, 115*(3), 672–682.

Chapter 9

Co-digestion Case Studies

Shirin Estahbanati, PhD, PE; Jay Surti; & Jack Sutton

1.0 INTRODUCTION TO CO-DIGESTION FACILITIES IN THE UNITED STATES

Water resource recovery facilities (WRRFs) can leverage co-digestion to enhance their operations on multiple fronts. By incorporating high-strength waste (HSW) into their anaerobic digestion processes, WRRFs stand to accrue numerous benefits. Co-digestion presents an opportunity for WRRFs to bolster their financial standing by increasing biogas production, potentially generating additional revenue streams, and reducing operating costs associated with waste management and energy consumption. Moreover, embracing co-digestion aligns with environmental and community interests, as it facilitates the diversion of organic waste from landfills and incinerators, thereby mitigating greenhouse gas emissions and improving air and water quality.

1.1 Brief Description of Co-digestion Facilities in the United States

Historically, the adoption of co-digestion in the United States can be traced back to pioneering efforts such as the construction of early manure digesters in the 1970s. These ventures, while not without challenges, laid the groundwork for understanding the potential of anaerobic digestion technology. Despite initial setbacks, subsequent advancements and global initiatives have driven the widespread adoption of co-digestion practices, fostering its integration into modern wastewater treatment strategies (BioCycle, 2009).

Sludge disposal and treatment remain significant challenges for WRRFs, exacerbated by escalating hauling costs and diminishing land availability due to tightening regulations on land application. Additionally, the incorporation of certain HSW streams into anaerobic digestion processes may necessitate regulatory approval from agencies outside the purview of municipal WRRFs, such as those governing food, agriculture, and environmental protection. Permit considerations are region or state specific and may include requirements for biosolids application, air quality, and inbound waste streams. Furthermore, once HSW is blended with conventional wastewater solids for co-digestion, the entire solids stream falls under the regulatory framework outlined in 40 CFR Part 503. Compliance with these standards, particularly regarding vector attraction requirements and parameters like volatile solids (VS) and chemical oxygen demand (COD) concentrations, adds to the operational complexity and regulatory burden faced by WRRFs. Moreover, increased quantities of digestate resulting from co-digestion may escalate handling and disposal costs, further straining operational resources (Ely & Rock, 2014).

In examining the policy and regulatory framework governing co-digestion facilities, it's essential to consider federal, state, and local laws, regulations, permits, and incentives that influence their establishment, operation, and expansion. Regulatory oversight, particularly under 40 CFR Part 503, governs the handling, reuse, or disposal of digested solids resulting from co-digestion processes. The incorporation of HSW streams introduces additional layers of regulation and permitting requirements, often involving agencies with diverse mandates beyond traditional wastewater management. These complexities increase the administrative burden on WRRFs, requiring dedicated resources for sampling, analysis, monitoring, and reporting. Moreover, as co-digestion practices evolve and volumes of digestate rise, the financial implications of compliance and waste handling must be carefully managed to ensure the sustainability of co-digestion initiatives within the existing regulatory framework.

This chapter explores the current market trends and growth prospects for co-digestion facilities in and outside of the United States and offers valuable insights drawn from real-world implementations. These insights arise from the practical experiences of WRRFs facing biosolids management challenges, including rising operating cost and meeting regulatory requirements. Through case studies, it becomes apparent that these challenges have driven WRRFs to investigate co-digestion as a viable solution, motivated by their renewable energy goals and environmental commitments. Furthermore, economic incentives serve as significant drivers for facilities to integrate HSW streams into anaerobic digestion processes. However, the implementation of HSW receiving stations requires careful consideration. Case studies highlight the importance of conducting comprehensive market assessments to evaluate disposal costs for haulers and the potential impacts of tipping fees on overall economic viability. While co-digestion initiatives may not always yield economic benefits, case studies highlight their potential to achieve other non-economic objectives, such as improved sustainability practices and increased energy production. These insights provide valuable guidance for navigating the operational complexities and economic implications of co-digestion projects within the context of real-world case studies.

In the landscape of case studies focused on co-digestion, the necessity of collaboration among key players and stakeholders emerges as a central theme. Within this realm, successful co-digestion projects are not only achieved through the efforts of a single entity, but rather through the collaboration of utilities, regulatory agencies, research institutions, vendors, and the public. This collaborative approach is important to navigate the complexities of regulations and environmental considerations in co-digestion initiatives. Regulatory agencies play a key role in establishing the necessary frameworks and standards, ensuring compliance with environmental regulations while fostering innovation. Utilities and municipalities contribute indispensable infrastructure and operational expertise essential for the efficient functioning of co-digestion facilities.

In the technological landscape of co-digestion facilities in the United States, a diverse array of HSW streams, including fats, oil, and grease (FOG), food industry waste, septage, and more, are co-digested with sewage sludge. Before implementation, thorough assessment of HSW's chemical composition is vital, encompassing parameters such as total solids, VS, COD, and nutrient content. This evaluation ensures alignment with facility objectives, meets pretreatment requirements, and identifies infrastructure needs while anticipating issues like odor and gas cleanup. By optimizing anaerobic digestion systems based on these assessments, facilities enhance operational efficiency and environmental performance, minimizing overall facility impacts.

1.2 Importance of Case Studies in Understanding Best Practices

In our case studies, we examine how co-digestion initiatives offer environmental and economic benefits. These include waste diversion, greenhouse gas reduction, renewable energy generation, soil improvement, and cost savings for municipalities and industries. Understanding the costs, benefits, and return on investment is crucial for project champions driving these initiatives. By exploring real-world implementations in this chapter, insight is provided on how co-digestion programs contribute to sustainable waste management practices, shaping the landscape of modern environmental initiatives.

2.0 SELECTION CRITERIA FOR CASE STUDIES

This section describes the selection criteria utilized to short-list co-digestion facilities to serve as case studies described herein:

Type of HSW: Co-digestion programs at wastewater utilities across North America have evolved around using different types of HSW streams. These HSW streams range from fats, FOG-type waste streams, and processed food waste streams ranging from pre-consumer type waste generated from industrial food manufacturing facilities to post-consumer waste streams through recovery of organics fraction of municipal solid waste. These waste streams possess unique characteristics ranging from pH, total solids, and volatile solids fraction, COD:VS ratios, odor profile, and so forth. Co-digestion facilities typically have implemented pre-processing systems to condition the waste stream for co-digestion. Case studies selected and described herein have been selected to capture design and operational knowledge and experience related to processing of diverse nature of waste streams.

Geographic location: Geographic location is a criterion considered in the selection of case studies and to capture the characteristics of communities and sub-regions that the wastewater utilities serve across the United States. A case study from Australia was also selected to draw global lessons learned. The purpose is to look at different geographic regions and take into consideration criteria that are also unique and representative of a sub-region, for example, climate, community makeup, that is, residential versus industrial versus a combination of residential and industrial, availability and accessibility of waste streams.

Regulations: Several states across the United States have adopted regulations regarding organic waste diversion from landfills, and regulations diverting organic waste produced by large-scale producers to beneficial use. For example, SB1383 is an effort to reduce organic waste in California landfills by 75% by 2025, and to increase recovery of excess edible food by 20% by 2025. Similarly, in 2022, the New York State Food Donation and Food Scraps Recycling Law went into effect, prohibiting landfill disposal of food and recycling inedible food through composting or anaerobic digestion. The Climate Leadership and Community Protection Act of 2019 also requires limiting statewide green house gas (GHG) emissions to 60% of 1990 levels by 2030 and to 15% by 2050. Impact of existing and emerging regulations on facilities that have adopted co-digestion has been investigated. Typically, WRRFs also have air permits to comply with. The air permit and compliance requirements range by state and region. Thus, case studies that represent varied air districts across the country have been selected to evaluate the impact of co-digestion.

Facility size: Co-digestion facilities across the United States vary in size from small- to medium-scale facilities (5 MGD to 30 MGD), to large-scale facilities greater than 30 MGD. Facilities selected as case studies and described herein are to capture the range of facility size, and how the size of the facility factored in the decision-making process to implement co-digestion. It is demonstrated that co-digestion programs can be successful and implemented for WRRFs that are diverse in size.

Wastewater treatment process and nutrient removal requirements: water resource recovery facilities are required to meet treatment standards as stipulated in the National Pollutant Discharge Elimination Systems (NPDES) Permits and goals established to achieve solids stabilization in accordance with the U.S. EPA Part 503 and state-level regulations. The NPDES permit requirements can range from meeting total suspended solids (TSS) and biochemical oxygen demand (BOD) reduction requirements, to achieving advanced level of treatment to reduce nutrients (e.g., nitrogen and phosphorus) in the final effluent. To achieve treatment requirements, WRRFs adopt a range of treatment processes such as activated sludge, biological nutrient removal (BNR) processes, and enhanced biological phosphorus removal (EBPR) processes. Further, WRRFs also use tertiary treatment processes such as filtration to enhance TSS and nutrient removal. The wastewater solids (sludge) produced during the liquid treatment process are treated using various stabilization processes to meet U.S. EPA Part 503 and state-level regulations. Co-digestion case studies selected herein represent a diverse range of liquid and solids treatment processes that have been adopted. Impact, if any, of co-digestion program on a WRRF's NPDES permit and biosolids beneficial use program is evaluated.

Biosolids beneficial use: As stated above, WRRFs use a diverse range of solids stabilization process and have biosolids beneficial use program in accordance with U.S. EPA Part 503 and state level regulations. Water resource recovery facilities that implement co-digestion have to be mindful about the impact of co-digestion on the solids treatment process and biosolids beneficial use. Thus, biosolids beneficial use is considered a criteria in the selection of case studies and to investigate the impact of co-digestion.

3.0 COMPARATIVE ANALYSIS OF MULTIPLE CASE STUDIES

In this section, a comparative analysis of six co-digestion facilities selected based on specific criteria is provided. These criteria include the type of HSW processed, geographic location, regulatory environment, facility size, wastewater treatment processes and nutrient removal requirements, and biosolids beneficial use programs. These facilities represent diverse approaches and practices in co-digestion across North America, including one facility outside of the United States.

These case studies, summarized in Table 9.1, offer lessons in the design, operation, and regulatory compliance of co-digestion facilities. They demonstrate that successful co-digestion programs can be implemented across a range of facility sizes, geographic locations, and regulatory environments, providing an understanding of the factors influencing co-digestion practices.

TABLE 9.1 Comparative Analysis of the Selected Case Studies

Name	Capacity	Nutrient Limitation	Type of HSW	Address
EBMUD	120 MGD (dry weather permitted capacity); 320 MGD primary capacity	Yes, monitoring requirements without specific effluent limits (Tentative Order for 2024)	Liquid organics, FOG, protein waste, food waste, other HSW	375 11th St, Oakland, CA 94607, USA
CMSA	7 MGD (average flow) 10 MGD (design capacity average flow)	No	Food waste, FOG	1301 Andersen Dr, San Rafael, CA 94901, USA
Gwinnett County	50 MGD (annual average day flow); 60 MGD (maximum month day flow)	Yes	FOG	1600 One Way, Buford, GA 30519
Oneida County	50 MGD (average flow)	Yes	Food waste	51 Leland Ave, Utica, NY 13502, USA
Stevens Point WTP	2.4 MGD average flow; 4.5 MGD permitted flow	Yes	Brewery wastes, food processing residuals	300 Bliss Ave, Stevens Point, WI 54481, USA
SAWater	60 MLD (design average dry weather flow)	Yes	Brewery wastes, FOG, whey, confectionary and food court waste	Anderson Avenue, Glenelg North SA 5045, Australia

4.0 CASE STUDY 1: EAST BAY MUNICIPAL UTILITY DISTRICT

The East Bay Municipal Utility District (EBMUD) was selected as a key case study for this chapter due to its exemplary and pioneering efforts in co-digestion. As one of the first facilities in the United States to implement a comprehensive co-digestion program, EBMUD has demonstrated success in managing high-strength organic waste (HSOW) and producing renewable energy. The facility's ability to integrate diverse feedstocks, such as food scraps and grease interceptor wastes, while maintaining environmental

TABLE 9.2 East Bay Municipal Utility District (Hyatt & Chakrabarti, 2024)

Facility Background	**Location:** Oakland, California, U.S. **Size:** 120 MGD (dry weather permitted capacity); 320 MGD primary capacity. Average dry weather flow is ~50 MGD **Type of wastewater process:** high-purity oxygen activated sludge. **NPDES Permit:** CA0037702 **Solid Handling Facility:** Anaerobic digestion (11 digesters): operates a two-stage digestion system, with the first-stage digesters at low thermophilic temperatures (50–52°C) and the second-stage digesters at conventional mesophilic temperatures (~37°C). This setup allows for enhanced breakdown in the first stage and further stabilization in the second. For digestor mixing, the first-stage digesters use draft tube mixers. The second-stage digesters are currently unmixed due to structural limitations but will receive pumped mixing within the next 6–9 months. Following digestion step is dewatering by centrifuges (5 centrifuges), and further biosolids treatment.
Co-digestion program	**Type of HSW:** Liquid Organics (50% of daily mass), FOG (20% of daily mass), Protein Waste (20% of daily mass) Food Waste (5% of daily mass), Other HSW (5%) **Operational History:** Since 2002
Technical information	**Receiving Stations:** EBMUD has three receiving stations. One receiving station is dedicated to low-strength wastes that are processed through the entire treatment plant. A high-strength receiving station, which was constructed in 2004, has five receiving tanks, two of which can accept solid food waste. A second high-strength waste receiving station was constructed in 2014 to allow off-loading outside of the treatment plant fence line. This station has five receiving bays and can only accommodate liquid waste. Waste from both high-strength receiving stations is pumped to a sludge blend tank prior to the digesters as described below. **Food Waste:** Source-separated food waste is processed and screened offsite and then tipped into a receiving station where it is mixed and slurried with process water and then further preprocessed as describe below. **HSW Processing System:** Primary sludge and thickened waste activated sludge are mixed with high-strength waste in a heated blend tank. Installation of the blend tank has helped with biogas equalization. Approximately 200,000 gallons of each, primary sludge, TWAS, and high-strength waste are blended and fed to the eight first-stage digesters.

(*continued*)

TABLE 9.2 East Bay Municipal Utility District (Hyatt & Chakrabarti, 2024) (*Continued*)

Technical information (*continued*)	**Current Practice for Odor Control:** Air Filtration, Treatment Process Improvements (EBMUD, 2017). **Consideration of Contaminants of Emerging Concern (PFAS, microplastics, other):** EBMUD has participated in various studies to characterize contaminants of emerging concerns, including PFAS and microplastics (Orner et al., 2022). EBMUD has piloted a sludge screening process, and a high-strength waste grit removal process, but has not implemented either at full-scale. **Sidestream Processing:** Currently no sidestream processing. Since 2021, EBMUD has been conducting a full-scale pilot study to achieve nitrification/denitrification within its pure-oxygen process. Implementing sidestream nutrient management technologies to meet more stringent discharge limits may also be considered (Orner et al., 2022).
Permitting requirements	**Nutrient Permit:** The 2014 Nutrient Watershed Permit entails effluent monitoring requirements without incorporating specific effluent limits (EBMUD, 2017). The Tentative Order, with a proposed effective date of October 2024 for NPDES CA0038873 Nutrient Watershed Permit would require significant reductions in total inorganic nitrogen in the future.
Operational and maintenance experience	**FOG:** EBMUD accepts a significant amount of FOG. These frequently come with a significant amount of contamination (rocks and gravel similar to concrete aggregate). This contamination breaks equipment and settles out in the digesters. **Food Waste:** EBMUD has been accepting small amounts (up to about 20 tons per day) of ground food waste since 2004. Effective preprocessing to remove contamination is key. Contaminants found in food waste include silverware, plastic bags, oyster shells, wine corks, rocks and grit, and other non-digestible items. These contaminants, especially film plastic, metals, and grit, can damage the onsite processing equipment, leading to significant maintenance costs and down time. Onsite preprocessing equipment for food waste: • **Grinder:** Vogelsang Rotacut has a rock trap and grinder. • **Hose Pump:** Bredel hose pump can manage high-grit content. • **Pulper:** The Brown paddle finisher is utilized to separate digestible from non-digestible material. **Biogas:** Beneficially reusing the biogas produced from HSW can become challenging as it can be significantly more variable with a high fraction of HSW in the feed. EBMUD experiences both a daily and a weekly fluctuation in biogas flow as HSW generators typically follow a normal working schedule (Monday to Friday).

(*continued*)

TABLE 9.2 East Bay Municipal Utility District (Hyatt & Chakrabarti, 2024) (*Continued*)

Financial	**Gross Revenue:** More than $16M per year
Biosolids beneficial use	All biosolids are beneficially reused through a combination of land application, composting, and thermal hydrolysis.
Any Innovarion	TruckTrack Software, an internally designed software package that efficiently monitors and processes all waste deliveries. This software controls security access, tracks delivery information, and automates billing processes, ensuring seamless operations (Jones et al., 2019). TWIST Software, an internally designed software package that monitors the impact of waste deliveries on plant operations.

and economic standards, makes it a model for sustainable waste management practices. East Bay Municipal Utility District's innovative approaches, including their technical capacity for research and pilot testing, careful permitting processes, and successful public and private partnerships, provide valuable insights into the potential benefits and challenges of co-digestion programs. Their experience and leadership in the field highlight the impact of co-digestion on modern environmental initiatives (Blate, 2022).

4.1 Factors Contributing to Success

The success of EBMUD co-digestion facility can be attributed to various factors across economic, environmental, and social domains. Economically, the facility has experienced substantial growth in revenue, particularly from tip fees associated with HSOW acceptance. With approximately 100–150 trucks of liquid waste received daily from 250+ different generators, EBMUD earns significant income from tip fees currently ranging from $0.04 to $0.11 per gallon for liquid wastes and $30 to $100 per ton for solid organic wastes (food waste). Annual revenue from tip fees has steadily grown since the inception of the co-digestion program in 2002, gross tip fees in fiscal year 2023 totaled more than $16 million.

Environmentally, the co-digestion facility plays a crucial role in waste diversion and renewable energy generation, contributing to GHG reduction and sustainable waste management. By accepting diverse feedstocks including food scraps, food and beverage manufacturing wastes, and grease interceptor wastes, these materials are kept out of landfills where they might otherwise result in methane emissions. Moreover, the facility utilizes biogas to produce onsite renewable heat and power that displaces grid electricity.

East Bay Municipal Utility District has partnered with public and private entities to pilot new technologies and develop projects for managing local food scraps. East Bay Municipal Utility District frequently presents at local, state, and national conferences to share its findings and lessons learned. East Bay Municipal Utility District also provides technical tours for both public and private entities from the United States and internationally to share its experiences and help advance and expand co-digestion. The success factors of the EBMUD can be summarized as:

- Significant excess digestion capacity.
- A culture that supports innovation to try new processes or changes in operations, such as pushing organic loading beyond textbook values.
- Technical capacity to conduct research and run pilot tests for the co-digestion of various materials.
- Careful permitting and monitoring of wastes permitted to discharge.

4.2 Managing Costs and Operations: EBMUD's Trucked Waste Program

One of the most important aspects of managing the costs of the trucked waste program is being careful about what wastes are permitted and accepted. Wastes that are high in contamination can cause significant and expensive problems while clean wastes can have little downside to acceptance. At EBMUD all wastes are required to be characterized, approved, and permitted before acceptance. All wastes haulers are also required to be permitted and all drivers must be trained. Requiring district staff involvement for deliveries can schedule appointments during regular business hours, ensuring efficient handling of waste materials.

East Bay Municipal Utility District is also careful about requirements of permit holders. Requirements include:

- Signing a base agreement that outlines permit holder responsibilities
- Insurance (general liability, auto, workers compensation)
- Onsite training of all trucked waste drivers by EBMUD staff

One way EBMUD keeps operational costs down is by charging trucked waste customers for a full load for every delivery, regardless of the actual volume. This avoids the need to carefully meter and track actual delivery volumes as well as any potential disputes.

East Bay Municipal Utility District's Resource Recovery Program has six dedicated staff members, including engineers, administrative personnel, one field inspector, and an environmental specialist. In addition, the program requires resources from throughout the wastewater department in operations, maintenance, engineering, and the laboratory. It was important to get buy-in from the entire organization when a new trucked waste program begins to make sure everyone understands both the potential benefits and the additional work required.

4.3 Additional Insights and Considerations

Hauling and Traffic Issues

East Bay Municipal Utility District is conveniently located at the intersection of several major freeways, so the additional 100–150 trucks that deliver hauled waste are an insignificant addition to the existing traffic. To further mitigate traffic impacts, a new high-strength receiving station was built outside of the facility fence line, reducing truck traffic within the treatment facility.

Permit Updates and Impact

The Regional Water Quality Control Board incorporated food waste receiving and treatment within the NPDES permit for EBMUD. Additionally, EBMUD obtained a rendering license through the California Department of Food and Agriculture to continue accepting animal blood waste. However, EBMUD discontinued the acceptance of blood waste during the dry season of 2023 to reduce nutrient loading and meet the goals of the Nutrient Watershed Permit (NPDES No. CA0038873).

Unspecified Challenges

The co-digestion facility faces several challenges, including contamination, grit, equipment damage, and added costs for operations and maintenance, as well as chemical inputs. Performing analytical analyses on unusual matrices that are not typical wastewater proves difficult. Additionally, tracking and monitoring wastes to ensure consistency with the permit description is crucial.

Facility Simulation Potential

Facilities with excess digestion capacity and high energy costs could potentially replicate EBMUD's co-digestion process successfully.

Lessons Learned

Key lessons learned from the operation of the co-digestion facility include:

- The facility and staff excel at handling liquids, but managing high-solids wastes can be challenging.
- Contamination control is a significant issue, and education alone is insufficient for managing source separated organics contamination.
- Careful permitting and monitoring of wastes are essential for effective operation (per interview with M. Hyatt and A. Chakrabarti of EBMUD in 2024).

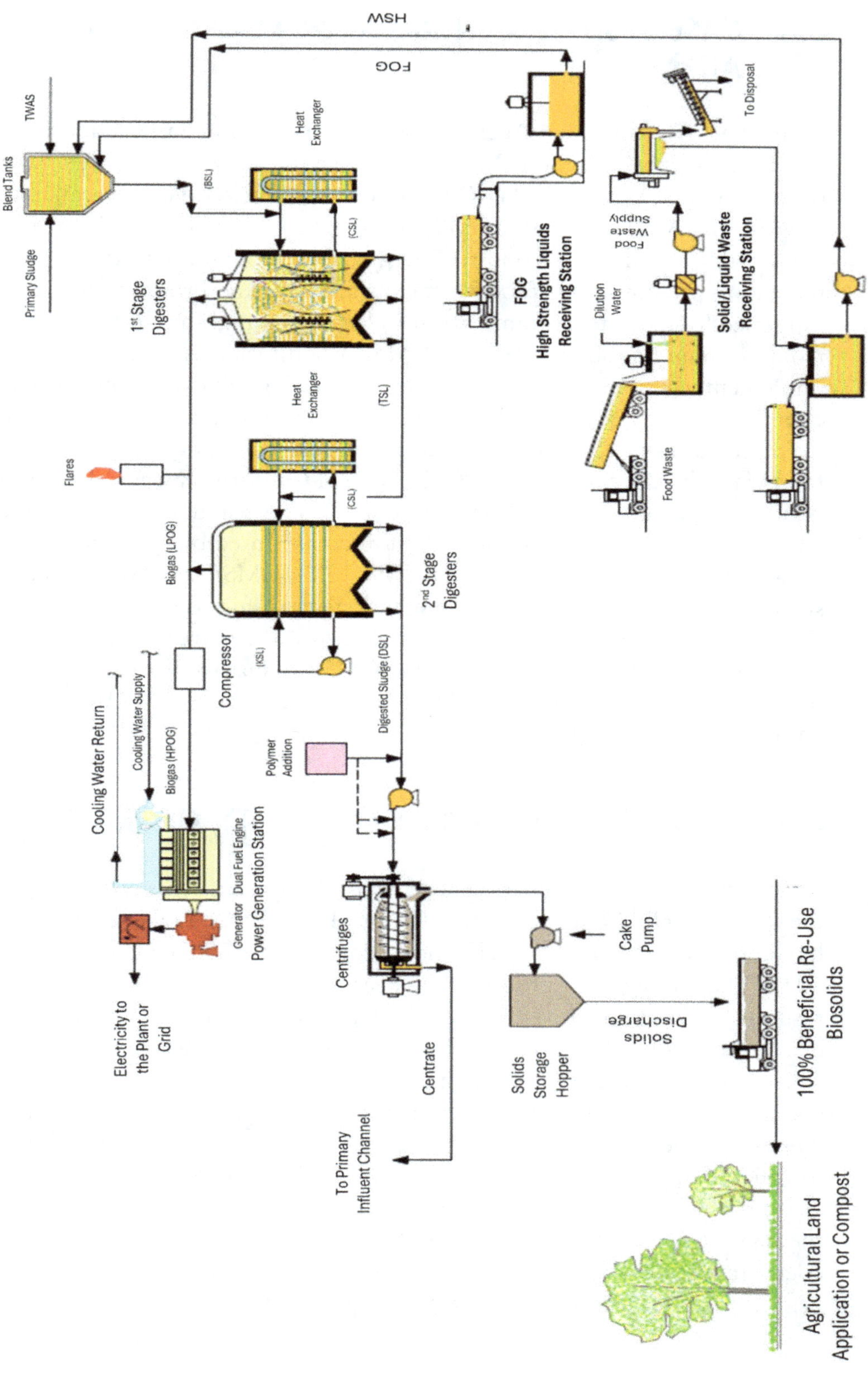

FIGURE 9.1 Co-digestion Process Flow Diagram at East Bay Municipal Utility District (Hyatt & Chakrabarti, 2024)

5.0 CASE STUDY 2: CENTRAL MARIN SANITATION AGENCY

Central Marin Sanitation Agency (CMSA) is located in San Rafael, California, and operates a WRRF with an average daily dry weather flow of 10 million gallons per day (MGD) and a peak wet-weather capacity flow of 125 MGD. In 2008, CMSA partnered with Marin Sanitary Services (MSS) to initiate a Food-to-Energy (F2E) program with the goal of managing pre-consumer food waste sustainably. The program aimed to minimize food waste contamination by accepting high-quality feedstock from commercial sources. Central Marin Sanitation Agency and MSS implemented quality control measures, including educating food waste contributors on contamination reduction techniques. The F2E program exemplified a proactive response to reach Marin County's goal of zero food waste by 2025, and supports the state of California's waste diversion regulation (Senate Bill 1383), which requires a 75% reduction in organic waste diverted to landfills and a 25% increase in edible food recovery. By aligning with the goals of Senate Bill 1383, CMSA and MSS demonstrated a commitment to environmental stewardship and sustainability. This process increased biogas production, producing renewable energy from a sustainable source and allowing CMSA to maximize the utilization of their existing assets (Surti, 2023). Figure 9.2 depicts CMSA's current co-digestion process flow diagram.

6.0 CASE STUDY 3: GWINNETT COUNTY, GEORGIA

The Gwinnett County Department of Water Resources owns and operates the F. Wayne Hill Water Resources Center (FWHWRC). Gwinnett County has implemented a FOG co-digestion program at FWHWRC since early 2012. The co-digestion program has been very successful for the county, resulting in additional revenue from tipping fees, and maximizing the use of existing anaerobic digestion capacity and enhancing digester gas generation. The digester gas is conditioned and used as a fuel in an internal combustion engine. Electricity generated on-site is used to meet a portion of the facility's electric demand and reduce dependence on grid electricity. In addition to generating electricity for FWHWRC's use, the cogeneration system has also produced sufficient heat to keep the anaerobic digesters operating in the mesophilic temperature (95–100 °F) range. Before implementation of cogeneration, digester gas was pressurized and used in boilers for generating heat for digester heating. Excess digester gas was flared.

TABLE 9.3 Central Marin Sanitation Agency

Facility Background	**Location:** San Rafael, California, United States **Size:** 10 MGD (dry weather permitted capacity) **Type of Wastewater Process:** Inlet screening, grit removal, primary sedimentation, high-rate trickling filters, aeration basins, secondary clarification, chlorination and dechlorination before discharge to the San Francisco Bay. **NPDES Permit:** CA0038628 (California Water Board, 2023) **Solids Handling Facility:** Solids processing at CMSA involves rotary drum thickeners for secondary sludge thickening, and two (2) mesophilic anaerobic digesters retrofitted with membrane covers for biogas storage. Thickened secondary sludge, primary sludge as well as HSW are the feedstocks used for the digesters. Anaerobically digested sludge is dewatered by centrifuges.
Co-digestion program	**Operational History:** In 2012, CMSA constructed the organic waste receiving facility to increase biogas production and throughput for the cogeneration engine. **Source Management:** Food waste at MSS undergoes metal removal and mechanical pretreatment before it is transported to CMSA. At CMSA, the food waste is mixed with fats, FOG from grease traps and interceptors in restaurants, markets, and similar types of food service establishments that are delivered by licensed haulers. In 2022, CMSA started accepting additional HSW materials such as pre-processed food wastes from other hauling companies, as well as de-packaged expired food waste from grocery stores and food processors, which are also first pre-processed at MSS. The resulting HSW undergoes further screening and mixing before being co-digested with wastewater sludge.
Technical information	**Facility Overview:** Initial construction included a 20 000-gallon working volume underground tank, a heavy object trap, chopper mixing pumps, a rock trap grinder, a paddle finisher, hose pumps to pump the contents to the digesters, and an odor control system. In 2024, CMSA upgraded the facility to include a 15 000-gallon cross-linked polyethylene aboveground receiving tank, a rotary lobe pump external to the receiving tank to provide mixing, and replacement of the heavy object trap with a Strainpress screening system. **Receiving Station:** The HSW receiving equipment is located outside, and the underground tank is primarily used to receive solid food waste material or liquid HSWs that are delivered by trucks without vacuum pumps. The food waste delivery is typically between 12%–24% solids and is off-loaded from end-dump trucks into a 10-ft by 8-ft double-leaf hatch in the roof of the underground tank. The aboveground tank can receive liquid food waste and/or FOG. The receiving equipment includes a delivery pad for unloading of HSW trucks. A camlock connection enables offloading of the liquid organic wastes from the trucks. The pad has a drain to the underground tank to contain any spills during unloading.

(continued)

TABLE 9.3 Central Marin Sanitation Agency (*Continued*)

Technical information (*continued*)	**HSW Processing System:** Liquid deliveries are screened by the Strainpress before being discharged into the 15 000-gal polyethylene receiving tank. Various flushing points are located along the pipelines to enable operators to clear out blockages. An actuated valve at the bottom of the tank allows a specified volume of tank contents to drain to the underground tank to mix with the solid food waste. Once offloading is complete and the level in the underground tank is above a minimum setpoint, the chopper mixing pumps blend the organic waste for about an hour. The mixed slurry is then pumped from the underground tank to the paddle finisher and rock trap grinder which removes small plastic pieces and any remaining debris before the slurry then goes into the screened sump. When the level in this sump rises to an adjustable setpoint, the organic waste feed pumps are initiated, and the slurry is pumped into the digesters in a dedicated pipeline over to the anaerobic digesters where it mixes with digested sludge before entering the digesters. **Current practice for odor control:** A fan extracts air from the organic waste receiving station headspace to be processed through a scrubber to treat and minimize odor. **Biogas:** The existing biogas treatment system consists of two (2) hydrogen sulfide removal vessels, gas compressors to pressurize biogas, a gas conditioning system that consists of two (2) chillers and a heat exchanger to remove moisture, and carbon media vessels for siloxane removal. Once pre-treated, the biogas is sent to a 995-kW combined heat and power (CHP) engine to generate electricity and heat. **Sidestream Processing:** The NPDES permit includes an average ammonia limit of 60 mg/L as nitrogen and maximum daily ammonia limit of 110 mg/L as nitrogen. There is no requirement for sidestream treatment of the dewatering centrate at this facility. A tentative order under NPDES Permit CA0038873 is anticipated to be adopted in July 2024, which would further regulate nutrient discharges to San Francisco Bay during dry weather.
Permitting requirements	**Nutrient Permit:** CMSA has not experienced NPDES permit compliance issues due to implementation of co-digestion. **Air Permit:** CMSA's organic waste receiving station is included in the facility's air permit. Additionally, the air permit limits the total material throughput through the organic waste receiving station as well as places limits on the emissions from the odor scrubber.

(*continued*)

TABLE 9.3 Central Marin Sanitation Agency (*Continued*)

Operational and maintenance experience	**Electricity Generation:** CMSA has more than a decade of operational and maintenance experience co-digesting food waste and generating biogas to produce electricity to achieve energy neutrality as well as exporting excess power back to the grid. **Contamination:** Managing contamination is key to operating a successful program. Contamination is managed in multiple layers, by control at the source through the MSS outreach program, by the preprocessing machinery at the MSS facility, and by the additional screening and grinding systems at the CMSA organic waste receiving station. Proper equipment selection for screening and pumping of these materials is also critical for success, as well as a robust preventative maintenance program. **Process Monitoring:** CMSA undertakes careful monitoring of various process parameters to maintain digester stability as CMSA does not have a redundant digester. This includes frequent and fine-grained monitoring of various process metrics with the goal of spotting any process instabilities as early as possible. One such method is the analysis of volatile fatty acids (VFAs). Central Marin Sanitation Agency noticed that traditional titration methods do not differentiate VFA species, which is necessary to delineate potential biological stresses on the digesters. Central Marin Sanitation Agency began the implementation of a VFA analysis method using a Metrohm Ion Chromatography (IC) system, which allowed CMSA to differentiate VFA species and streamline their analysis (per interview with Koekemoer and Carbajal of CMSA in 2024). Additionally, CMSA has successfully used a real-time biological sensor, Sentry, on the digester recirculation lines since 2020, which alerts operators to potential digester instability issues. Central Marin Sanitation Agency routinely tracks a list of digester health metrics including Sentry signal, VFA levels, pH, and biogas production to ensure digester stability. Central Marin Sanitation Agency has also developed highly valuable rules of thumb for the digester external organic waste loading rates, which are strictly monitored to ensure stability. This includes the ratio of FOG loading to other external organic waste loading such as food waste. **Inspection Program:** CMSA has a sampling and inspection program to ensure that external organic waste deliveries meet minimum quality standards.

(*continued*)

TABLE 9.3 Central Marin Sanitation Agency (*Continued*)

Financial	**Capital Costs:** CMSA invested $2 million for the initial organic waste receiving station and $7.65 million in digester upgrades to install new flexible membrane covers, pump mixing systems, and hydrogen sulfide scrubbers (Water Environment Federation [WEF] & Environmental Law Institute, 2021). Central Marin Sanitation Agency spent roughly $3M from 2022–2024 to upgrade the HSW receiving station with a new screening system, 15 000-gal storage tank, mixing pump, pressure washer, offloading pad, as well as a new heat exchanger and chillers for biogas processing; this project was funded in part by CalRecycle's Organics Grant Program. An upsized cogeneration system was constructed for $7 million in 2022 that helps use the additional biogas generated from food waste. Additional expenses were incurred to provide the ability to export excess power to the grid. It is important to note that not all of the above costs are directly attributable to the organic waste program. The financial considerations of a co-digestion program are highly complex and include a combination of "sunk" costs (such as digester mixing systems, membrane covers or engines) that CMSA could incur regardless of co-digestion due to aging infrastructure and other reasons, and additional costs such as the organic waste receiving facility that are only due to accepting external organic waste streams. Additionally, CMSA also had significant excess digester capacity from its original facility construction and CMSA received significant grant funding to pay for portions of its new organic waste receiving facility, gas treatment system, and upgraded cogeneration system. **Operational Costs:** There are significant costs to operate the receiving station, which include periodic contaminant removal, heavy equipment wear and tear due to the low pH and abrasive feedstocks, and increased laboratory sampling costs. Other operating costs such as digester or engine operation include costs that would be incurred regardless of acceptance of organic waste, so the specific incremental costs attributable to organic waste do not lend themselves to easy calculation. **Revenue:** CMSA facility power costs are very high and typically increase each year, more than general inflation. Therefore, offsetting facility power costs with self-generated bioenergy has a significant financial benefit. In 2024, CMSA consistently reached energy neutrality with long periods of being 10%–30% energy positive—exporting excess bioenergy to the local grid. The excess bioenergy is purchased by a local utility company, Marin Clean Energy, under a 10-year power purchase agreement providing additional revenue to CMSA. By generating a substantial portion of its electricity onsite, through the use of biogas, CMSA reduced its reliance on external energy sources and realized considerable cost savings over time (WEF and Environmental Law Institute, 2021).

(*continued*)

TABLE 9.3 Central Marin Sanitation Agency (*Continued*)

Financial (*continued*)	The cogeneration system also provides 100% of CMSA's significant on-site heating demands. Lastly, CMSA charges MSS and other FOG and organic waste haulers a tipping fee for each gallon delivered to CMSA, presenting an additional revenue stream for the program. **Collaboration with MSS:** It is also important to note that the MSS ratepayer base overlaps with the CMSA ratepayer base and this successful co-digestion program helps both agencies in meeting their strategic objectives, a true "win-win" situation. Additionally, many local ratepayers also purchase power from Marin Clean Energy, which may include the locally generated power from CMSA.
Biosolids beneficial use	Class B biosolids are sent to Redwood Landfill for use as alternative daily cover, to land application through Synagro during dry weather, and a portion are sent to Lystek in Fairfield-Suisun for production of fertilizers that are injected into designated agricultural land in Solano County (WEF and Environmental Law Institute, 2021).

The primary drivers for the FOG co-digestion and cogeneration program for the county were purely economic as described below:

Declining revenues due to the economic downturn of 2007, resulting in significant budget concerns for the county.

The FWHWRC also processes wastewater solids received from the county's other wastewater treatment facility called the Yellow River Water Resources Center (WRC). Solids generated by the Yellow River WRC are discharged into an interceptor sewer and received with the influent wastewater at FWHWRC. This centralized treatment approach also contributes to enhancing biogas generation from anaerobic digesters.

The county completed a business case evaluation (BCE) during the planning process that concluded that a gas-to-energy facility based on ICEs was feasible. The BCE evaluated several different scenarios for generating power from biogas, including with and without FOG and HSW. The BCE recommended a gas-to-energy facility of approximately 2 MW to make the project financially viable. However, at the time FWHWRC was operating at approximately 50% of its total design capacity, and purchase of natural gas would have been required to obtain full power generation and heat recovery benefits. This condition further motivated FWHWRC to evaluate FOG and HSW co-digestion to increase digester gas production, minimize purchase of natural gas, and maximize use of anaerobic digestion and cogeneration capacity.

The county also conducted bench scale tests to evaluate the increase in biogas generation due to the addition of FOG to the anaerobic digesters.

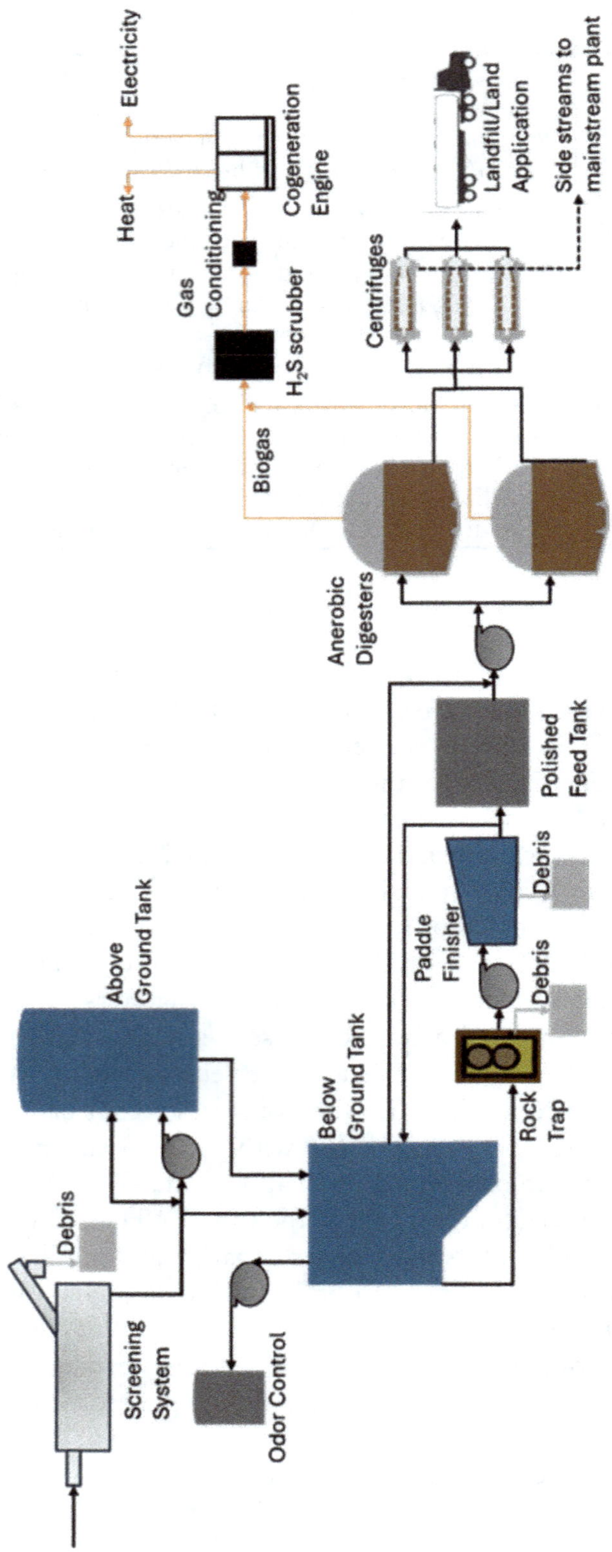

FIGURE 9.2 Co-digestion Process Flow Diagram at Central Marin Sanitation Agency

TABLE 9.4 Gwinnett County F. Wayne Hill Water Resources Center (Forbes et al., 2010; Van Horne et al., 2011)

Facility Background	**Location:** 1600 One Way, Buford, GA 30519 **Rated Capacity:** 50 MGD monthly average, 62.5 MGD monthly average **Liquid Treatment:** Preliminary treatment, primary clarification, activated sludge—A2O process, secondary clarification, tertiary filtration, pre-ozonation, biologically active filtration, final disinfection w/ ozone. **Solids Treatment:** Primary and secondary sludge holding for p-release, sludge co-thickening, mesophilic anaerobic digestion, final centrifugation, landfill cake disposal, nutrient recovery as struvite. The digesters utilize a draft tube mixing system. Each digester has a mixer mounted into a steel tube welded within the digester. The mixers are reversible but are typically operated in a downward flow. Each mixer has a 20 HP motor and is designed for 720 rpm and a minimum mixing flow of 6500 gpm. NPDES Permit No.: GA0038130
Co-digestion program	Composed of FOG and high-strength organic waste, operational February 2012.
Technical information	Pumped discharge from trucks into 80 000 gal of storage Permit requires pre-screening from haulers, dual mesh screens installed on discharge line, cleaned by truck driver after discharge. Direct batch feed to active digesters **2023 total received volume:** 12.9 MG Waste quality and toxicity screenings performed quarterly.
Permitting requirements	Biosolids are landfilled. The gas-to-energy facility operates under an air permit.
Operational and maintenance experience	Housekeeping needed for FOG discharge area multiple times per week. Rotary lobe pump maintenance is largest maintenance cost. Poor experience using hose pumps for mixing and dosing due to frequent hose breaks.
Financial	**Tipping Fee:** Base cost is $0.06/gallon; cost increases with high nutrient loads and inert solids. **2023 Income:** $881,000 **2023 Maintenance Costs:** $282,000 **Value of Biogas Generated:** Potential revenue and reduction in cost due to biogas beneficial use
Biosolids beneficial use	Currently landfilled, nutrient recovery as struvite of digested liquid fraction and thickened sludge liquid fraction in Ostara process. Biosolids drying to be implemented in next 5 years.

These tests help determine the quantity of FOG that would be required to be added to the anaerobic digesters each day to generate sufficient gas to fully utilize the cogeneration system capacity.

The county implemented the overall project as two separate design–build projects:

- A gas-to-energy project that included the digester gas treatment elements and the cogeneration system
  - The duration of the project from design through construction completion was approximately 20 months. The facility was built for a total cost of $5.5 million. The project was funded primarily through a state loan. A small portion of the cost was funded by county funding sources.
- An FOG and HSW project that included the receiving facilities, transfer pumping, storage tanks, and conveyance facilities to the existing anaerobic digesters.
  - The duration of the project from design through construction was approximately 13 months. The total cost of the project as $2.3 million. The project was funded through a combination of Department of Energy (DOE) grant funding and American Reinvestment and Recovery Act (ARRA) funding.

For the FOG and HSW receiving and storage facility, the county optimized the costs by repurposing and existing slab (reduce excavation and concrete costs), use reuse hose pumps from the WRC to provide unloading and digester feed capabilities. To address maintenance requirements, the receiving facility was designed with a hot water washdown system utilizing an in-line hot water heater as well as a containment area sized to hold rainfall and one full truck load of FOG/HSW. Other features of the facility include a rock trap with a flanged opening for debris removal and bypass piping, glass lined ductile iron piping, duty and standby piping to the anaerobic digesters, ability to bypass the tube-in-tube heat exchangers, ability to heat the storage facilities to approximately 43 °C (110 °F), and redundant tank level sensors in each tank to prevent overfilling due to any foaming that might occur.

A gas treatment system was installed to condition the digester gas before use as fuel in the new CHP system. The gas treatment provides hydrogen sulfide, siloxanes, and moisture removal. Hydrogen sulfide and moisture can cause corrosion of the cogeneration engine components and reduce the efficiency of the combustion process. Siloxanes can combust to produce fine silicon oxides. Deposition of silicon oxides inside the engine can result in

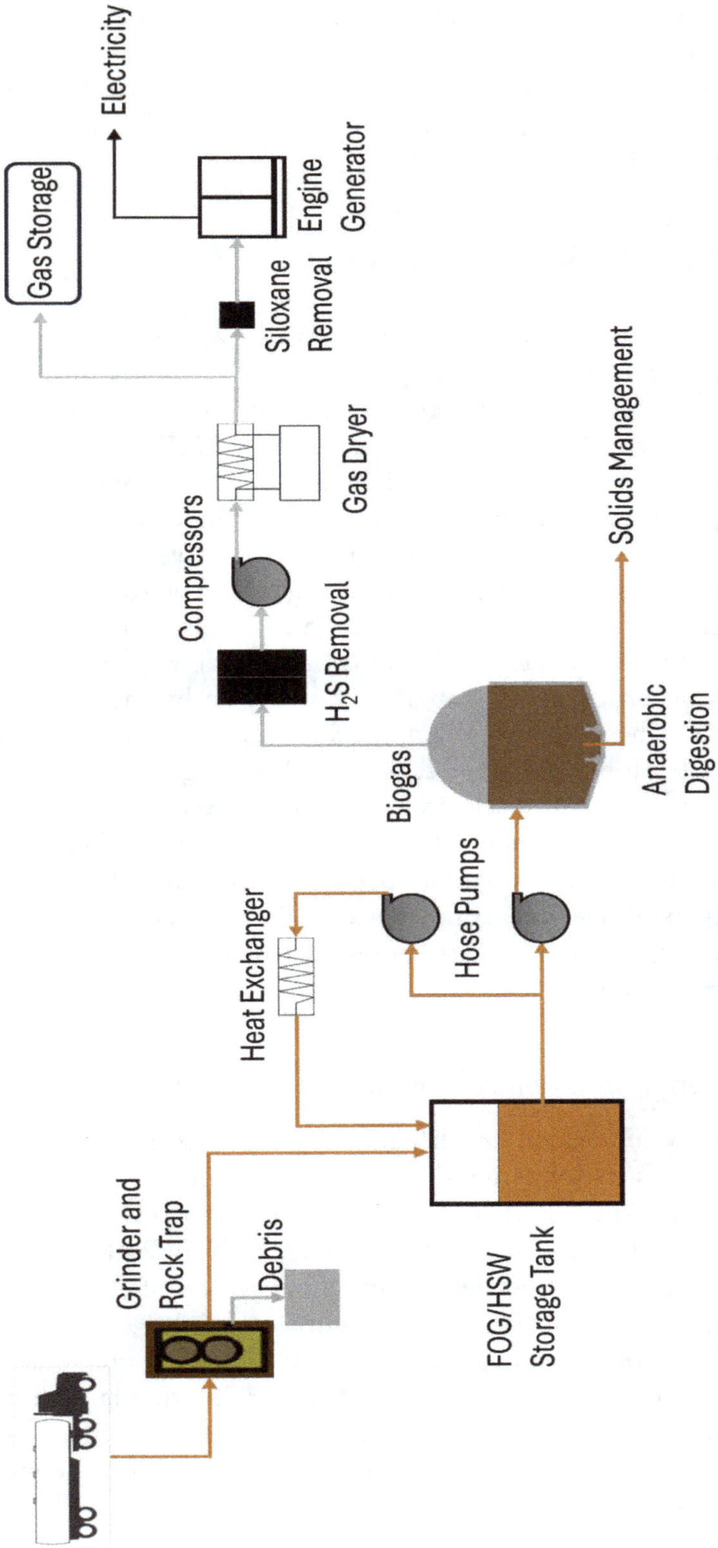

FIGURE 9.3 Co-digestion Process Flow Diagram at the Department of Water Resources

erosion of the engine components, increased maintenance, wear-and-tear, and premature failure.

The following is a process flow schematic of the HSW/FOG receiving, storage, feeding, digester gas treatment and cogeneration facility (Forbes et al., 2010; Van Horne et al., 2011).

7.0 CASE STUDY 4: ONEIDA COUNTY

Fed by 45 miles of interceptor sewers assisted by two pumping stations, the Oneida County Sewer District's WRRF, built in the 1960s, sits on a constrained site along the Mohawk River and serves 14 municipalities, including the city of Utica. With an average flow of 32 MGD, the WRRF was designed to handle 55 MGD. As population and development increased, the treatment capacity was insufficient to handle high wet-weather flows.

In response to sanitary and combined sewer overflows, the county entered into a consent order agreement with the New York State Department of Environmental Conservation (NYSDEC) in 2011 that required a doubling of capacity, to 111 MGD. A major upgrade was planned through a $300 million program. The upgrade was driven by the consent order, knowledge that the system was aging, and NYSDEC's requirement that the city of Utica develop a combined sewer overflow long-term control plan as part of its state pollutant discharge elimination system permit. An innovative split-flow treatment system and other upgrades have doubled facility capacity and reduced sanitary sewer overflow by nearly 90%. The upgrade also included extensive public outreach and education activities and formation of a steering committee with representatives from the sewer district's member municipalities.

7.1 The Upgraded Solids Handling Infrastructure

This case study focuses on another major component of the project: significant upgrades to solids handling facilities to build a new anaerobic digester system at the WRRF and a partnership with the Oneida-Herkimer Solid Waste Authority, whose facility is next to the WRRF. These efforts were undertaken to address regulatory changes that affect management of the county's solid waste and to maximize use of the wastewater treatment infrastructure. In early 2022, the New York State Food Donation and Food Scraps Recycling Law went into effect, prohibiting landfill disposal of food waste from large generators but allowing the alternatives of donating still-edible excess food and recycling inedible food through composting

TABLE 9.5 Oneida County Water Pollution Control Plant

Facility Background	**Location:** On Sewage Plant Road near 51 Leland Avenue, Utica, NY 13502 (nearby address) **Rated Capacity:** 65 MGD (dry weather treatment capacity); 111 MGD (wet weather treatment capacity) **Liquid Treatment:** Preliminary treatment (screening and grit removal), primary clarification with primary sludge degritting, activated sludge, secondary clarification, chlorine disinfection **Solids Treatment:** Thickening of degritted primary sludge with gravity thickeners, gravity belt thickening of waste activated sludge, two egg shaped primary mesophilic anaerobic digesters with draft tube mixers, secondary anaerobic digester with a pumped mix system, belt filter press dewatering of digested sludge. NPDES Permit No.: NY0025780
Co-digestion program	Co-digestion of processed food waste slurry received from the adjacent Oneida-Herkimer Solid Waste Facility.
Technical information	**Food Waste Slurry Production at the Oneida-Herkimer Solid Waste Facility:** Food waste is received on a tipping floor. From the Tip Floor, the food waste is transferred to a hopper that feeds a turbo separator. Food waste slurry is collected in two 25 000 gal holding tanks before being pumped to the WRRF. A Scott Equipment Turbo Separator Model T30 is utilized for depackaging, particle size reduction, and emulsification. Packaging material and contamination are removed from the Turbo Separator apart from the organic material. **Food Waste Slurry Treatment at the Oneida County Water Pollution Control Plant:** The slurry pumped from the solids waste facility is screened and de-gritted by a screening equipment called The Beast manufactured by SAVECO. **Biogas Conditioning and Beneficial Use:** The biogas is conditioned for hydrogen sulfide and siloxanes removal, dried to reduce moisture and compressed before being used as fuel in Capstone C600S microturbine based CHP system with hot water recovery.
Permitting requirements	Dewatered biosolids are landfilled. The gas-to-energy facility operates under an air permit.
Operational and maintenance experience	Housekeeping needed for The Beast at the Water Pollution Control Plant.
Financial	The public–public partnership operates under a mutually beneficial agreement. Dewatered biosolids are conveyed to the Oneida Herkimer Solid Waste Authority Landfill for disposal.

(*continued*)

TABLE 9.5 Oneida County Water Pollution Control Plant (*Continued*)

Financial (*continued*)	The average electrical demand at the Oneida County WRRF is 1400 kW. Implementing anaerobic digestion and co-digesting wastewater solids and food waste, and beneficially using biogas for cogeneration, has helped the county meet nearly 75% of their average electric demand through on-site power generation. Additionally, exhaust heat recovered in the CHP system exceeds 500 000 BTU/hour, which is adequate to heat the digester contents as well as three buildings at the facility.
Biosolids beneficial use	Currently landfilled

or AD. The Climate Leadership and Community Protection Act of 2019 also requires limiting statewide GHG emissions to 60% of 1990 levels by 2030 and to 15% by 2050. A favorable regulatory climate and the WRRF's proximity to the solid waste facility opened up an opportunity for a unique public–public partnership to help keep organics out of the landfill and comply with new laws.

As the food-waste recycling and GHG-reduction regulations were being promulgated, the solids treatment process at the WRRF was also undergoing a major upgrade. Central components were installation of new AD infrastructure to process wastewater solids and produce biogas, as well as a new cogeneration system that could use the biogas for fuel. The county decided to size the new AD infrastructure to allow co-digestion of food waste and wastewater solids, thereby keeping food waste out of the landfill and using it to produce biogas.

The Oneida-Herkimer Solid Waste Facility was already equipped to receive and process large generators' food waste through its F2E Program. Through this program, the organics that make up approximately 22% of the solid waste facility's waste stream are diverted. The source separated organics (SSO) from the solid waste facility, including the food waste from large generators, are now processed into a slurry and sent to the WRRF. The major upgrades at the WRRF, in turn, allow AD of the food waste slurry along with the WRRF's wastewater solids. The digester gas resulting from this co-digestion is then used to produce energy to heat and power the facility, reducing GHG emissions and increasing energy independence.

Figure 9.4 illustrates co-processing of food waste and wastewater solids. Food waste from large generators is received at the solid waste facility. A de-packaging process produces a food-waste slurry, which is stored in holding tanks before it is pumped directly through a forcemain to the AD facility at the WRRF. The slurry is screened and de-gritted to remove inorganic fractions such as sand, plastic, glass, and metals before it is fed to the

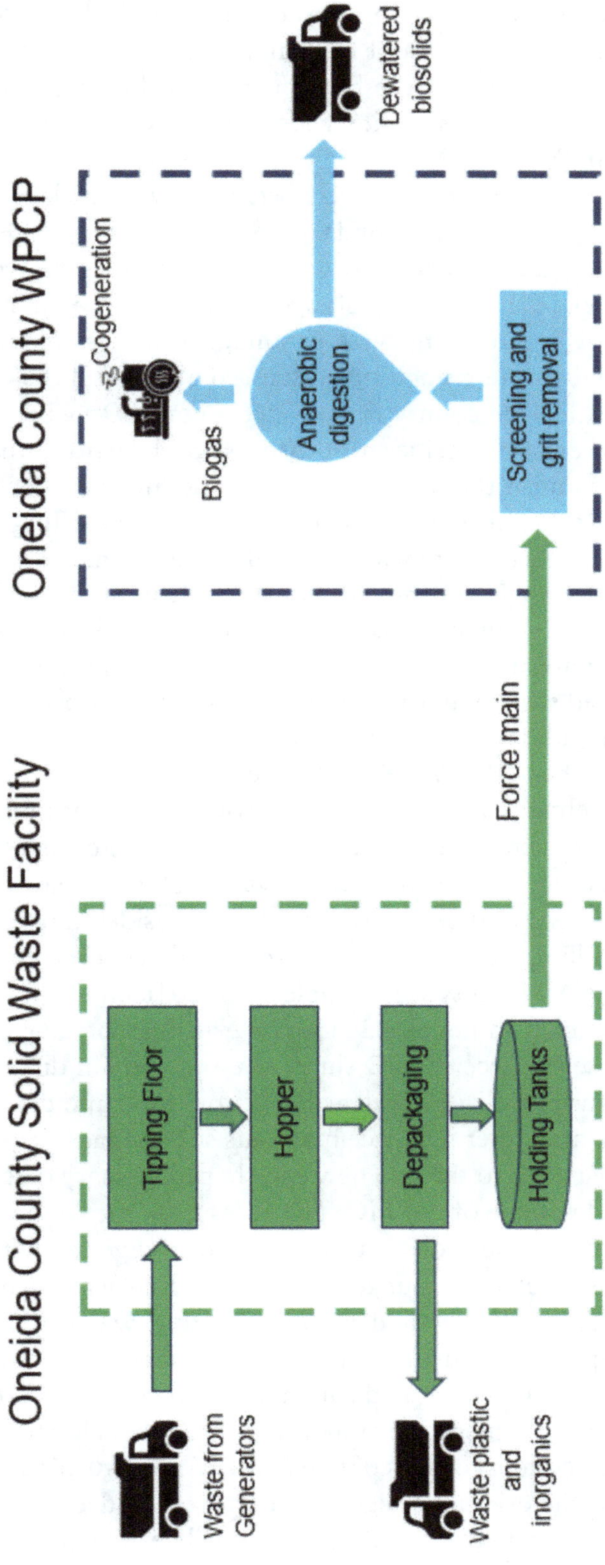

FIGURE 9.4 Co-digestion Process Flow Diagram at Oneida County (Surti, 2023)

anaerobic digesters. The biogas produced is used as fuel in a cogeneration system to produce electricity and to heat water. The electricity is used to meet the WRRF's needs, and the hot water is used to heat the anaerobic digesters and other buildings on site. Digested solids are dewatered and sent to the county landfill.

Before the WRRF upgrade projects began, gravity thickening was used for thickening of primary and secondary sludge. Belt filter presses dewatered the thickened sludge, and three fluidized bed incinerators treated the wastewater solids. In 2016, new sewage sludge incinerator maximum achievable control technology regulations were promulgated. These regulations, plus the emerging solid waste regulations discussed above and the county's sustainability goals, led the county to consider a new AD facility rather than to refurbish the existing incineration process and upgrade the emissions control system to meet the new regulations. The life-cycle cost of upgrading the existing incineration process and replacing the existing incineration process with a new AD facility was analyzed. In addition, the state identified energy recovery from AD systems as a key component of the Mohawk Valley Regional Sustainability Plan. Considering the results of the life-cycle analysis and other non-financial factors such as ease of operations and maintenance (O&M), enhanced sustainability, support for local food processing and dairy industries, and regulatory favorability, the county chose AD as the long-term solution to meet the WRRF's treatment needs.

Process modeling was used to predict solids quantities and characteristics, simulate predicted system performance, and determine the basis of design for upgrading treatment of wastewater solids. As noted above, the design of the new AD and dewatering facilities considered including excess capacity for handling increased volume from co-digestion with food waste.

Design of the AD facility was completed in 2016, and construction was substantially complete in January 2019. The existing septage-receiving facility was repurposed to receive SSO slurry via a forcemain directly from the adjacent solid waste facility, as well as FOG from food and dairy industries. Grit, inorganics, and other nuisance materials such as rags are screened out before the slurry is sent to the two new egg-shaped primary digesters, which have a combined volume of 2.4 MG. The SSOs from the solid waste facility and FOG from dairy and industrial sources are co-digested with thickened primary and waste-activated sludge from the WRRF's wastewater. The new secondary digester stores digested sludge prior to dewatering.

The biogas produced through AD is conditioned to remove moisture, particulates, hydrogen sulfide, and siloxanes, then compressed to meet fuel specifications for a cogeneration system that produces electricity and heat. The cogeneration system consists of five 200-kW microturbine generators—three included in the original design, and two more added to eliminate flaring of the excess biogas that resulted from co-digesting wastewater solids and food waste. Together, the five microturbines can produce up to 1000 kW of electricity, meeting up to 50% of the WRRF's needs, thereby reducing dependence on the grid. Heat recovered from the cogeneration process is used to heat the ADs and the buildings, thereby offsetting use of natural gas.

This case study demonstrates the benefits of a public–public partnership between two sister agencies (wastewater and solid waste) to co-create solutions to meet regulations, reduce carbon footprints, promote a circular economy, protect public health and the environment, and benefit ratepayers (Surti, 2023).

8.0 CASE STUDY 5: STEVENS POINT WASTE TREATMENT PLANT

The city of Stevens Point, Wisconsin, has transformed its wastewater treatment facility into a wastewater resource recovery facility to enhance the efficiency and effectiveness of wastewater treatment. The journey toward energy efficiency began in the early 2000s, as the facility became an early participant in Wisconsin's Focus on Energy program. Recognizing the potential for energy self-sufficiency, the facility invested in food waste feedstocks, an upgraded AD system, and a 180-kW CHP system. The CHP system operated for 10 years before being removed from service due to higher-than-expected maintenance costs. Today, the facility utilized biogas produced from co-digestion for heat production and biosolids drying, showcasing its leadership in sustainable wastewater management.

A critical component of this transformation is the HSW receiving station, established in 2014 through a partnership with Stevens Point Brewery. The station includes a 40 000-gal receiving tank equipped with a pumped mixing system with chopper-type centrifugal pumps to ensure consistent quality and quantity of feedstocks. Preprocessing equipment such as a bar screen and rock trap remove large debris and contaminants, protecting downstream equipment and improving processing efficiency. These measures ensure that the facility can handle a variety of organic wastes efficiently. Stevens Point Water Treatment Plant (SPWTP) started accepting HSW in 2012 utilizing two repurposed 6000-gal tanks as a way to prove a positive return on investment (ROI) on a HSW receiving station.

The anaerobic digesters at SPWTP play a vital role in breaking down organic matter, producing biogas and digestate. In 2018, the opportunity was presented to increase anaerobic digestion capacity when a biosolids dryer was added eliminating the need for a 1.6 MG storage tank. This additional digestion capacity allowed for a 100% increase in HSW receiving. The facility has also upgraded its existing digesters to enhance capacity and efficiency, accommodating the increased load from HSW feedstocks.

The biogas produced is utilized to produce heat for 7 buildings, 2.2 MG of anaerobic digestion, and biosolids drying.

Feedstock management at SPWTP focuses on high-strength wastes from local food industries and the Stevens Point Brewery. Initially accepting FOG, the facility now prioritizes more consistent and higher-quality filter performance rating to maximize biogas production and minimize operational issues. The partnership with Stevens Point Brewery is a key element, as the brewery's high-BOD wastes are piped directly to the facility's digesters, reducing energy costs and operational upsets associated with aeration.

In 2018, SPWTP broke ground on a new biosolids drying facility featuring a thermal oil paddle dryer. This has enabled the production of Class A biosolids, which is distributed as an agricultural fertilizer and topsoil additive. This addressed regulatory changes for nutrient treatment and market challenges for land application. The success of SPWTP's co-digestion program is attributed to incremental improvements, strong leadership, and strategic partnerships, serving as a practical model for other facilities aiming to enhance energy efficiency and resource recovery (Jones et al., 2019).

8.1 Factors Contributing to Success

After communication with the staff at the facility, operations staff commitment and communication were identified as the primary factors of success. Operations staff buy-in is considered the largest contributor to the success of the co-digestion program. Co-digestion at a high level requires full buy-in from the staff. When issues arise, it is essential that the staff communicates the information quickly and accurately so changes can be made to avoid digester upsets.

8.2 Managing Costs and Operations

Managing costs and operations at the WRRF involves careful consideration of various financial and operational factors.

Capital Costs

The co-digestion facility at the WRRF was established through a public–private partnership with the Stevens Point Brewery, with costs split 50/50 for the entire project. This included an 1800-gal stainless tank with pumps at the brewery, a 3300-ft forcemain, and a 40 000-gal HSW tank with a mixing pump and digester feed pump at the wastewater treatment facility. The total cost to the WRRF was less than $600,000.

Operational and Maintenance Costs

The operational costs for solids handling are less than $0.02 per gallon. Minimal costs are associated with co-digestion if there is excess digester

TABLE 9.6 Stevens Point Waste Treatment Plant (Interview with employees on the co-digestion process at Stevens Point WTP, 2024)

Facility Background	**Location:** Stevens Point, Wisconsin, United States **Size:** 2.4 MGD average flow; 4.5 MGD permitted flow **Type of Wastewater Process:** secondary design flow **NPDES Permit:** No adjustments were required to the plant's NPDES, air or solids permits **Liquid Treatment:** secondary treatment facility with an anoxic-oxic system for biological phosphorus removal **Solids Handling Facility:** The solids handling facility includes four primary mesophilic anaerobic digesters totaling 2.2 MG, with a 1.6 MG covered storage tank, a belt filter press, and a biogas-fueled paddle dryer. These digesters include two 45-ft diameter units with recirculation for heating as the only mixing, one 55-ft diameter unit with a linear motion mixer, and one 100-ft diameter unit with pump and nozzle mixing. This system processes all facility solids as well as high-strength wastes from a receiving station equipped with a pump and nozzle mixing system, bar screen, and rock trap. The paddle dryer produces Class A biosolids for distribution.
Co-digestion program	**Type of HSW:** Brewery wastes (6.5%), food processing residuals (93.5%) **Operational History:** Since 2012
Technical information	**Preprocessing Updates:** On-site 1-in. bar screen, rock trap, large grit sump at bottom of mixing tank, chopper pump and metered into digester. **HSW Processing System:** The HSW processing system at SPWTP includes a receiving station with a 40 000-gal tank, a pumped mixing system with chopper type centrifugal pumps bar screen, and rock trap. It uses a dedicated pipeline from the brewery and accepts hauled food waste, ensuring efficient preprocessing and digestion. **Current Practice for Odor Control:** Dropping FOG as a feedstock and pre-mixing the feedstocks before introducing them to the digester. **Pre-processing:** Pumped mixing system with chopper type centrifugal pumps, bar screen, and rock trap. **Consideration of Contaminants of Emerging Concern (PFAS, microplastics, other):** This is one of several upcoming concerns; because feedstocks are all pre-consumer food wastes, currently no specific issue was observed. **Sidestream Processing:** A small amount of ferric chloride is dosed to our belt filter press (BFP) filtrate to prevent struvite formation.
Permitting requirements	**Nutrient Permit:** Maintained good results staying within current WPDES permit so no extra permitting has been required.

(*continued*)

TABLE 9.6 Stevens Point Waste Treatment Plant (Interview with employees on the co-digestion process at Stevens Point WTP, 2024) (*Continued*)

Operational and maintenance experience	The operation and maintenance of the co-digestion system at SPWTP involve regular monitoring of feedstock quality, maintaining the high-strength waste receiving station, and ensuring the digesters function efficiently. Staff manage consistent feedstock supply, prevent operational issues, and ensure effective co-digestion processes.
Financial	**Capital Cost:** <$600,000 for the HSW feed system. **Annualized Operational Costs:** Minimal increase in electrical costs due to the 15hp mixing pump and 10 hp digester feed pump being added. **Annualized Maintenance Costs:** 2 hours/week of labor for cleaning the rock trap and replacement wear parts for the pumps every 18–24 months. No additional maintenance costs were incurred for operating the anaerobic digesters. **Annualized Revenue From Extra Biogas:** The WRRF saves $60,000 per year in heating costs from utilizing biogas for onsite. **Annualized Tipping Fees:** $400,000 **Annualized Contaminant and Inert Disposal Cost:** $0
Biosolids beneficial use	At the SPWPT biosolids are heat dried using a paddle dryer. These solids are utilized as agricultural fertilizer and by the city to enhance the topsoil they use in restoration of green spaces after construction projects.
Any Innovation	By diverting brewery wastes from the headworks, the WRRF saves $150,000 annually in wastewater treatment costs while being able to avoid a multi-million dollar facility upgrade. This pipeline also benefits the brewery, saving them $150,000 each year in wastewater surcharges. The SPWTP utilizes a dual fuel thermal oil boiler to heat and dry biosolids to Class A specifications using primarily biogas produced through co-digestion.

capacity. Regular maintenance includes 2 hours per week of labor for cleaning the rock trap and replacement of wear parts for the pumps every 18–24 months. No additional maintenance costs were incurred for operating the anaerobic digesters.

Economic Impacts of Sidestream Processes

The only impact of the sidestream process at SPWTP is the addition of extra ferric chloride to the BFP filtrate. This cost is included in the $0.02 per gallon of operational and maintenance costs. The ammonia limit is pH-based; if the pH is under 7.2, the limit is >59 mg/L, allowing operation without increased aeration needs.

Lessons Learned and Recommendations

This section summarizes key insights and guidance derived from various co-digestion programs. It addresses essential considerations for design, O&M, permitting, impacts, odor control, monitoring requirements, biogas recovery, utilization of additional biogas, and economic benefits. The goal is to provide a comprehensive framework for planning and operating future co-digestion facilities by referencing specific case studies.

8.3 Additional Insights and Considerations

Hauling and Traffic Issues

Very few issues have been encountered with trucking and traffic management at the co-digestion facility. The implementation of effective traffic management strategies has ensured smooth operations and minimal disruptions.

Permit Updates and Impact

The permit for the facility covers effluent and biosolids. It is essential that any accepted wastes do not jeopardize meeting these requirements. Detailed information about current permits, including the NPDES permit, is available to ensure compliance and effective management of the co-digestion process.

Unspecified Challenges

Anaerobic digestion is usually a portion of the facility that does not receive much attention due to the "set it and forget it" mentality. Accepting outside wastes requires the operations staff to understand how the biology works and how different feedstocks and feed rates react with the biology. Accepting outside wastes introduces a level of uncertainty to a very stable process, necessitating careful monitoring and management.

Facility Simulation Potential

Any facility with excess anaerobic digestion capacity can potentially replicate the co-digestion process. The primary clarifier scum tank was initially used as a receiving tank, with the material fed directly to the digesters to prove a positive ROI on an automated system. Once biogas production levels were established, effective utilization strategies were developed.

Lessons Learned

Significant lessons have been learned from the operation of the co-digestion facility. It is crucial to start slowly and vet all wastes before acceptance. The biggest issues have arisen from slug-loading feedstocks into the digesters, often due to feedstock changes without notification. Fats, oil, and gas, for example, varies significantly in COD and total solids, highlighting the need

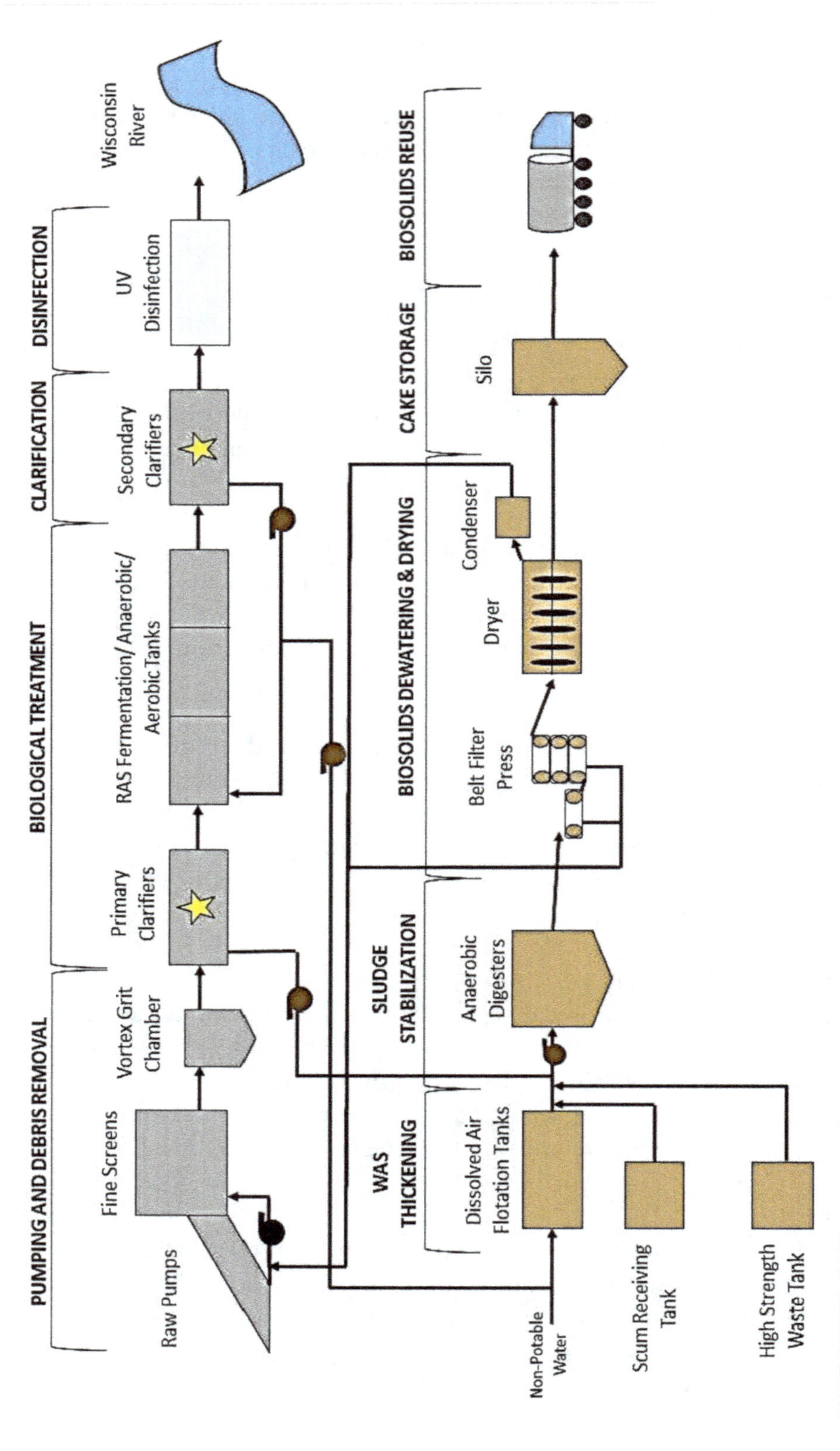

FIGURE 9.5 Wastewater Treatment Process Flow Diagram at Stevens Point Waste Treatment Plant

for consistent monitoring and management (Interview with Chris Lefebvre on the co-digestion process at Stevens Point WTP, 2024).

9.0 CASE STUDY 6: SAWATER, GLENELG WWTP CO-DIGESTION SCHEME

Glenelg WWTP is owned and operated by South Australia Water Corporation. The facility has a design capacity of 60 MLD. In 2012–2013, the co-digestion portion of the facility was constructed. This was the first of its kind in Australia. This case study offers operational lessons learn and demonstrates how sourcing high-strength wastes can translate to an increase in biogas production and electricity generation. Figure 9.6 below details the process flow diagram for the Glenelg co-digestion scheme.

10.0 SUMMARIZE KEY TAKEAWAYS FROM THE CASE STUDIES

EBMUD:

- Permitting and Monitoring: Ensure careful permitting and waste monitoring to comply with regulations.
- Partnerships for Innovation: Partner with public and private entities to pilot new technologies.
- Revenue Generation: Charge for full-load deliveries to manage operational costs effectively.
- Internal Software Solutions: Implement software packages for operational efficiency. Useful for large-scale facilities with complex operations.
- Separation of Waste Sources: Install different receiving stations to cater for the different types of HSW received, which may be of variable contamination and quality.

CMSA Food-to-Energy Program:

- Quality Control Measures and Education: Implement rigorous quality control measures and educate contributors on contamination reduction.
- Multiple Layers of Contamination Control: Implement pre-screening of HSW at multiple levels to reduce the likelihood of contaminated wastes damaging assets and impacting process performance.

TABLE 9.7 Glenelg Wastewater Treatment Plant (Forbes et al., 2010; Van Horne et al., 2011)

Facility Background	Location, Size of the Facility (WRRF–MGD), type of wastewater treatment process: **Location:** Glenelg North, South Australia, Australia **Size:** 60 MLD (design average dry weather flow) servicing 200,000 people. **Type of Wastewater Process:** Inlet screens, primary sedimentation, an integrated fixed film activated sludge secondary treatment facility, secondary clarification, ultra filtration, ultraviolet, and post chlorination for reuse of class A recycled water. **Permit:** EPA license 1560 (license includes three other metropolitan wastewater treatment facilities) (South Australian Water Corporation, 2014). **Solids Handling Facility:** The solids treatment process at Glenelg consists of five mesophilic anerobic digesters. The digesters are fed primary sludge, gravity thickened secondary sludge, as well as high-strength organics. Two of the digesters have floating covers storing for biogas, the other three digesters have fixed roofs. The digesters are mixed hydraulically by sludge recirculation pumps.
Co-digestion program	SAWater instigated a research project involving co-digestion of industrial high-strength organics to boost biogas production in the anaerobic digestion process. Successful trials led to the construction of a fully automated co-digestion facility at the Glenelg WWTP in 2012–2013. High-strength organic wastes co-digested at Glenelg WWTP include brewery, wine, grease trap, whey, confectionary, and food court waste (Dreyfus et al., n.d.).
Technical information	**Receiving Station:** To unload waste, truck operators access the site with a security key card. The waste is unloaded in a bunded area where the operator connects the tanker to a dedicated discharge camlock fitting. The tanker contents gravitate to an underground receival pit. Positive displacement/self-priming pumps deliver the waste from the receival pit to one of two storage tanks (31 kL volume each). The system is sized to receive one full load from a single 25 kL tanker. Due to the variability of contamination levels in the waste received, the strainers were prone to blockages. In order to suspend the waste and avoid accumulation of solids at the bottom of the tank, a dedicated recirculation pump draws waste from the base and returns it to the top of the tank on a periodic cycle. The blended material from the recirculation pump discharges to an adjacent thickened sludge storage tank, where mixing with the municipal sludge occurs before delivery to digester. A mesh is installed in the gutter to enable runoff from the bunded area to enter the receival pit.

(continued)

TABLE 9.7 Glenelg Wastewater Treatment Plant (Forbes et al., 2010; Van Horne et al., 2011) (*Continued*)

Technical information (*continued*)	**Current Practice for Odor Control:** No odor extraction or control exists at the site. **Biogas:** The anaerobic digesters at Glenelg WWTP are maintained at a sludge retention time of 20 days and produces biogas. The biogas is converted by three combustion engines for on-site electricity generation. Heat generated from cogeneration is used to heat water that consequently heats recirculated digested sludge in a heat exchanger. There is also a waste gas burner present.
Permitting requirements	No noted impacts on discharge compliance were associated with the scheme.
Operational and maintenance experience	Some operational lessons learnt from the Glenelg co-digestion scheme are outlined below. **Maintenance:** Operations are required to manage pipe blockages, increase pump maintenance, and conduct frequent tank inspections. **Process Monitoring:** pH monitoring of the co-digestion substrates is required to avoid any shock loads (Dreyfus et al., n.d.). **Heat Tracing:** The installation has involved polyvinyl chloride (PVC) pipe from the pump to the storage tanks, which is not heat traced. Heat tracing considered in future SAWater co-digestion projects (Chen et al., 2017). **Supervisory Control and Data Acquisition (SCADA) Control of Loading Rates:** The Glenelg WWTP SCADA system was developed to allow operators to determine different substrate loading rates that are applicable based on the substrate received (Dreyfus et al., n.d.). **H_2S Concentrations:** H_2S concentrations increased from 1816 ppm to 2137 ppm following the implementation of co-digestion (Dreyfus et al., n.d.).
Financial	**Capital Costs:** The project was implemented as part of the capital delivery budget from 2009. The capital cost was minimized by the use of high-density polyethylene tanks, positive displacement pumps, and PVC piping. **Revenue:** No tipping fee is obtained for the waste received at the facility. Before co-digestion, the anaerobic digestion process and electricity generation via cogeneration produced enough renewable energy to meet 55% of the facility's on-site usage. Since implementing co-digestion, the facility is able to meet 65%–85% of its on-site electricity usage from the additional biogas generated through co-digestion (Dreyfus et al., n.d.).

(*continued*)

TABLE 9.7 Glenelg Wastewater Treatment Plant (Forbes et al., 2010; Van Horne et al., 2011) (*Continued*)

Biosolids beneficial use	Anaerobically digested biosolids are dewatered, stockpiled, and then taken off-site to be used as a soil conditioner in the production of crops in South Australia. The biosolids are required to meet regulatory requirements under the Environmental Protection Act (1993) and are distributed free of charge to farmers (Dreyfus et al., n.d.).

- Facility Upgrades: Regularly upgrade facilities to maintain efficiency and revenue benefits.
- Process Monitoring: Routinely monitor digester health and process stability through a variety of key metrics, that is, VFAs, pH, sentry signal, biogas production, organic loading rate, and ratio of organics to municipal solids in co-digestion.

Gwinnett County:

- Project Drivers: Rising energy and operating costs, economic downturn of 2007, and declining revenue motivated the county to investigate FOG and HSW co-digestion.
- Business Case and Sizing the FOG/HSW Co-digestion Program: The county conducted a BCE to determine the economic viability of a new biogas fueled cogeneration system. The county estimated the increase in biogas production that would be required to maximize the use of the cogeneration system. An estimate of additional biogas required to fully utilize the cogeneration system, coupled with bench-scale tests conducted to estimate biogas generation per unit mass of FOG/HSW, helped the county estimate the daily quantity of FOG/HSW required for co-digestion. This evaluation further informed the size of the new FOG/HSW receiving and storage facility.
- Funding and Project Delivery: The project was funded through a combination of DOE grant funding and ARRA funding. The facility improvements were completed through the design–build alternative delivery mechanism.
- Program Financial Benefits: The county co-digestion program is matured and has been operating for more than a decade. The tipping fees from receipt of FOG and HSW has been a major revenue generator for the county over the years. In addition, the county has been able to produce renewable electricity to reduce dependence of grid electricity.

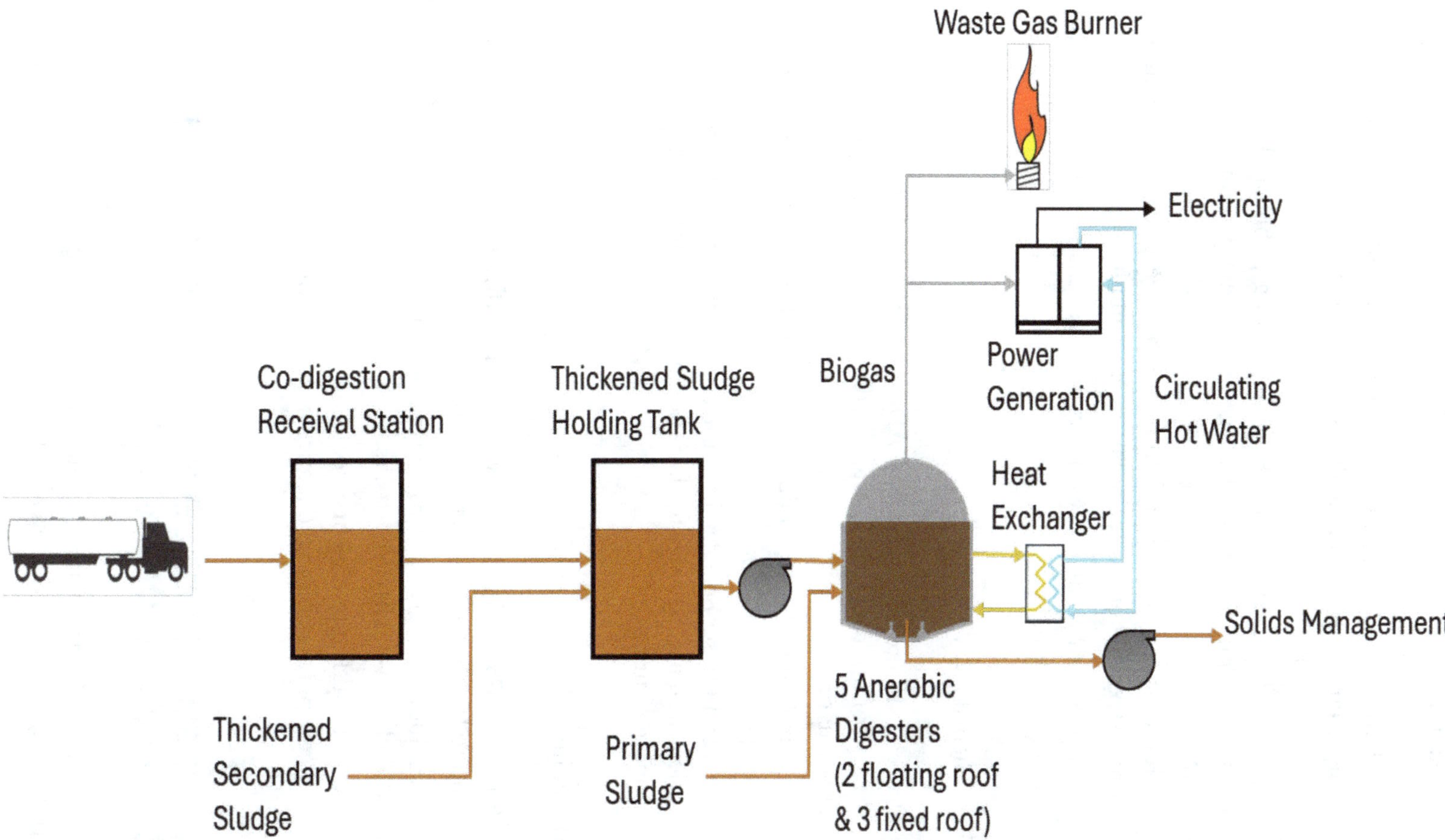

FIGURE 9.6 Co-digestion Process Flow Diagram at Glenelg Wastewater Treatment Plant

Oneida County:

- Regulatory Compliance and Proactivity: Respond proactively to regulatory changes. Ideal for facilities in regions with evolving environmental regulations.
- Public–Private Partnerships: Establish strong public–public partnerships with municipal waste management entities to enhance program sustainability.
- Community Engagement: Conduct extensive public outreach and form a steering committee to ensure community support.
- Life-cycle Cost Analysis: Conduct comprehensive cost analyses for decision-making.

SAWater Glenelg WWTP:

- Process Monitoring: Enable SCADA systems to assist operators with the use of appropriate digester loading rates depending on the type of waste received.
- Regular Maintenance: Maintain equipment regularly to ensure reliable operation.
- Beneficial Reuse: Beneficially reuse biogas and distribute biosolids for agricultural use. Suitable for facilities looking to enhance their circular economy initiatives.

Stevens Point WTP:

- Energy Self-Sufficiency: Achieve near self-sufficiency in energy by integrating high-strength waste and utilizing biogas for cogeneration.
- Local Industry Partnerships: Partner with local industries for consistent and high-quality feedstocks.
- Incremental Improvements: Focus on incremental improvements and strategic partnerships to enhance program efficiency.

11.0 POSSIBILITY OF DEVELOPING A DATABASE FOR CO-DIGESTION FACILITIES

In order to provide information for utilities considering co-digestion at their facility, having access to a comprehensive database would be a good resource.

Developing a comprehensive database for co-digestion facilities can provide numerous benefits:

- Data Sharing: Facilitate data sharing among facilities to improve design and operational strategies (e.g., insights from EBMUD's software solutions).
- Benchmarking: Enable benchmarking and performance comparisons to identify areas for improvement (e.g., CMSA's quality control measures).
- Research Support: Support research and development efforts by providing access to real-world data (e.g., SAWater Glenelg WWTP's pilot testing results).
- Regulatory Compliance: Enhance regulatory compliance by documenting and sharing successful permitting and monitoring approaches (e.g., Oneida County's proactive regulatory response).

12.0 COST/REVENUE, BIOSOLIDS IMPACTS, DEWATERING IMPACT, GAS USAGE, ETC.

- Cost/Revenue:
  - Tipping Fees: Generate significant revenue from tipping fees for high-strength waste acceptance (e.g., EBMUD and Stevens Point WTP).
  - Energy Savings: Achieve energy savings by utilizing biogas for on-site power generation (e.g., Stevens Point WTP) or generate excess electricity and supply renewable energy to the power grid (e.g., CMSA).
- Gas Usage:
  - Renewable Energy Generation: Utilize biogas for onsite energy generation to reduce reliance on external energy sources (e.g., CMSA and Stevens Point WTP).

13.0 RECOMMENDATIONS FOR FUTURE CO-DIGESTION FACILITY PLANNING AND OPERATIONS

Design and Infrastructure:

- Advanced Technologies: Invest in advanced preprocessing and treatment technologies to enhance biogas production and operational efficiency (e.g., EBMUD).

- Facility Flexibility: Design facilities with flexibility to handle various feedstocks and incorporate innovative technologies (e.g., Stevens Point WTP).

Operations and Maintenance:

- Maintenance Schedules: Implement regular maintenance schedules to ensure reliable operation (e.g., SAWater Glenelg WWTP).
- SCADA Upgrades: Incorporate SCADA upgrades to assist operators in managing organic loading rates to digesters (e.g., SAWater Glenelg WWTP)

Regulatory and Community Engagement:

- Regulatory Compliance: Ensure compliance with all regulatory requirements and maintain open communication with regulatory bodies (e.g., Oneida County).
- Community Support: Engage with the community and local industries to build support and secure partnerships for feedstock supply and program sustainability technologies (e.g., Stevens Point WTP).

Economic and Environmental Impact:

- Revenue Generation: Maximize economic benefits by generating revenue from tipping fees and utilizing biogas for renewable energy (e.g., EBMUD).
- Environmental Sustainability: Promote environmental sustainability by reducing GHG emissions, diverting organic waste from landfills, and producing high-quality biosolids for agricultural use (e.g., CMSA and SAWater Glenelg WTP).

By integrating these lessons learned and recommendations from specific case studies, future co-digestion facilities can achieve enhanced efficiency, sustainability, and economic viability.

14.0 REFERENCES

BioCycle. (2009). Historical perspective: Farm digesters. *BioCycle*. https://www.biocycle.net/historical-perspective-farm-digesters/

Blate, M. (2022). WRF Project# 5307: Evaluation of source separated organic feedstock pretreatment and management practices. *Proceedings of the Water Environment Federation's Technical Exhibition and Conference*.

California Water Board. (2023). *NPDES Permit CA0038628*.

Chen, M., Wishart, I., & Laginestra, M. (2017). *C5268—Bolivar WWTP co-digestion facility—Concept design report.* South Australia Water Corporation [Preprint].

Dreyfus, J., Kroff, P., & van den Akker, B. (n.d.). *Boosting energy production through co-digestion.* https://www.waternz.org.nz/Attachment?Action=Download&Attachment_id=3399

East Bay Municipal Utility District. (2017). *Reducing nutrients in the San Francisco Bay through additional wastewater treatment plant sidestream treatment.* East Bay Municipal Utility District and Project Partners.

Ely, C., & Rock, S. (2014). *Food waste to energy: How six water resource recovery facilities are boosting biogas production and the bottom line.* [Preprint].

Forbes, B., Jalla, S., & Horton, J. (2010). Fast track to new digester gas-to-energy facilities at the F. Wayne Hill water resources center, Gwinnett County, Georgia. *Proceedings of the Residuals and Biosolids Conference*, pp. 882–893. Water Environment Federation.

Jones, C. A., Coker, C., Kirk, K., & Reynolds, L. (2019). *Food waste co-digestion at water resource recovery facilities: Business case analysis.* The Water Research Foundation [Preprint].

Koekemoer & Carbajal. (2024). Interview with employees on the co-digestion process at Central Marin Sanitation Agency.

Orner, K. D., Smith, S., Nordahl, S., Chakrabarti, A., Breunig, H., Scown, C. D., Leverenz, H., Nelson, K. L., & Horvath, A. (2022). Environmental and economic impacts of managing nutrients in digestate derived from sewage sludge and high-strength organic waste. *Environmental Science & Technology*, *56*(23), 17256–17265.

South Australian Water Corporation. (2014). *EPA Licence 1560: Environmental Authorisation under Part 6 of the Environment Protection Act 1993.* Retrieved July 14, 2024, from https://www.epa.sa.gov.au/licence_docs/EPA1560%20-%20South%20Australian%20Water%20Corporation%20-%20Glenelg%20North.pdf

Surti, J. (2023). The opportunity for water resource and recovery facilities to advance regional GHG reduction goals. *Proceedings of the Residuals and Biosolids Conference.* Water Environment Federation.

Van Horne, M. P. , Bruton, T., Hardy, S., Bullard, C. M., & Long, H. (2011). From grease to green: Two case studies of FOG receiving, co-digestion and combined heat and power generation. *Proceedings of the Water Environment Federation, 2011*(6), 516–532.

Water Environment Federation & Environmental Law Institute. (2021). *Food waste co-digestion at Central Marin Sanitation Agency (CA): Business case analysis case study.*

Conclusion

Co-digestion of organics such as food waste, FOG and similar feedstocks is one of the few carbon-negative and proven approaches for beneficially reusing these materials that allows recovery of renewable energy as well as nutrients such as carbon, nitrogen and phosphorus. Anaerobic digestion utilizes microbes in the absence of oxygen to break down organic matter to the final products of methane and carbon dioxide (biogas) as well as ammonia, sulfur, phosphorus, and bicarbonate. The biogas can be utilized to produce renewable energy, and the other constituents (e.g., C, N, P) that remain within the digestate can be beneficially used as a soil amendment.

The main drivers for co-digestion programs that are discussed in Chapters 1 and 2 include:

- regulatory drivers: local and state regulations requiring diversion of organics from landfills
- economic drivers: revenue through tipping fees and energy generation
- sustainability goals: utilities working toward energy and/or carbon neutrality or more sustainable approaches to address waste issues

Aligning a co-digestion program with the goals of a WRRF (sustainability, economics, regulations) is an important step in planning and development of a co-digestion program, as is stakeholder involvement. Stakeholder engagement is critical, and stakeholders could include operators, local communities affected by the program, public officials, management, engineering, and laboratory staff, as discussed in Chapter 2. Planning a co-digestion program should include evaluations on regulations such as air quality requirements, NPDES permits, biosolids end use requirements, and permits and requirements for waste haulers. A business case evaluation (BCE) is essential to demonstrate economic viability of the project. The BCE should include capital costs, operational costs, and overall life-cycle costs. While this may present itself as a routine exercise, the nuances and risks that need to be considered to develop a robust financial model include conducting a sensitivity analysis around factors such as tipping fees, volume of material received, and biogas production and utilization and impacts on dewatering and recycle streams. Program management should also be evaluated when planning a co-digestion program, which should include operation and maintenance duties and staffing, monitoring, sampling, laboratory testing, and recordkeeping.

Consideration of different feedstock is vital to achieving the goals of a co-digestion program and is discussed in detail within Chapter 3. The types, sources, and characteristics of feedstocks for co-digestion are numerous, including food waste, FOG, animal wastes and processing residuals, agricultural wastes, brewery, and winery wastes. Typical parameters used for characterizing different feedstocks include TS and VS, COD, N and P content, as well elemental composition including metals. Ultimately, the different feedstocks can be evaluated by their three main components, proteins, lipids, and carbohydrates which have different impacts on the digestate chemistry and biogas production and content as discussed in more detail in Chapter 4. BMP and bench-scale testing can help determine the potential for biogas production and digestate chemistry impacts.

Stoichiometry is a very useful tool that can help estimate the potential impacts of different types of feedstock on biogas quantity and quality (CH_4, CO_2 content), and digestate chemistry (pH, alkalinity, ammonia) and characteristics such as TS and VS. In addition, downstream impacts such as conditioning and dewatering can be impacted along with the final dewatered cake quality. For example, in many cases, polymer demand increased with co-digestion but the wet solids leaving a facility did not increase until higher loading rates were used. Similarly, dewatering sidestreams can be impacted such as with higher ammonia and phosphorus loads, which can impact operations of the liquid treatment processes from these recycle flows. The application of stoichiometry for co-digestion along with the potential for downstream impacts are detailed in Chapter 4.

Depending on their source, feedstocks can have varying composition and characteristics with the potential for contaminants that can impact the digestion process. A number of preprocessing approaches can be used to pretreat and improve the consistency of feedstocks and remove contaminants. These processes can include biological as well as mechanical processes and are discussed in detail in Chapter 5.

A number of design considerations should be evaluated for implementing a co-digestion program. Paramount among these considerations are regulatory requirements such as NPDES and air permitting as well as biosolids end use. In addition, the receiving facilities, existing digester capacity, and biogas handling system must be evaluated along with the mixing, pumping, and dewatering processes as discussed in Chapter 6.

Successful implementation of a co-digestion program ultimately relies on skilled personnel and requires appropriate staffing. In addition, creation of SOPs, training programs, and troubleshooting guides are needed to support the day-to-day operations. The details for developing these tools are discussed in Chapter 7.

The increased production of biogas is often one of the main drivers for considering a co-digestion program. Biogas, comprising mainly CH_4 and CO_2, can be used for myriad purposes including production of heat, combined heat and power, injected into the natural gas pipeline as renewable natural gas, and/or used to fuel vehicles. These options can require gas cleanup to remove impurities such as siloxanes and hydrogen sulfide, and in the case of RNG the CO_2 would also need to be removed. All of these options reduce fossil fuel use and, as a result, reduce the carbon footprint of a facility. These considerations of biogas quantity and usage as well as the environmental benefits are discussed in more detail in Chapter 8.

A number of WRRFs have significant experience with implementation of co-digestion programs that can provide useful insights on the benefits and challenges that arise. Chapter 9 reviews several case studies of WRRFs that implemented co-digestion. For example, EBMUD successfully partnered with public and private entities to ensure regulatory compliance and has grown revenues total more than $16 million in 2023 from their co-digestion program. They have had to implement processes for contaminant challenges and they have experimented with diverse feedstock through research and piloting different feedstock, creating a model co-digestion program. Chapter 9 discusses five additional case studies on co-digestion highlighting the successes and lessons learned from each WRRF.

Index

www.ingramcontent.com/pod-product-compliance
Lightning Source LLC
LaVergne TN
LVHW082118210925
821603LV00003B/25
* 9 7 8 1 5 7 2 7 8 4 8 1 9 *